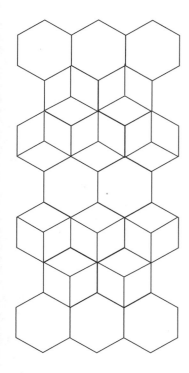

Investigating Patterns

Symmetry and Tessellations

Jill Britton

Dale Seymour Publications®

Dedicated to ToniAnn Guadagnoli, editor par excellence, and
Mark Veldhuysen of Cordon Art.

A special thank you to student artists: Len Church, Steve Dawson, Henry
Furmanowicz, Lyda Kobylansky, Stephen Makris, Tanya McLain, and
Elizabeth Oliviera.

Managing Editor: *Catherine Anderson*
Project Editor: *ToniAnn Guadagnoli*
Project Sponsor: *Merle Silverman*
Production/Manufacturing Director: *Janet Yearian*
Production/Manufacturing Coordinator: *Roxanne Knoll*
Design Director: *Phyllis Aycock*
Cover and Interior Designer: *Monika Popowitz*
Graphics: *Jill Britton*

This book is published by Dale Seymour Publications®,
an imprint of Addison Wesley Longman, Inc.

Dale Seymour Publications
10 Bank Street
White Plains, NY 10602-5026
Customer Service: 800-872-1100

Printed in the United States of America
Order number DS21886
ISBN 0-7690-0083-5

1 2 3 4 5 6 7 8 9–MZ–03 02 01 00 99

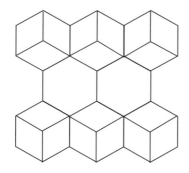

CONTENTS

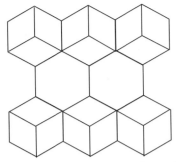

ACKNOWLEDGMENTS

"Mirror" and "upside down" graphics, pages 12 and 13, ©1981 Scott Kim. All rights reserved. www.scottkim.com

First Nations art, page 21, used with permission from the artist, Art Thompson.

Logos, page 27 (listed left to right)

Row 1 The encircled VW logo is a registered tradmark of Volkswagen of America, Inc.
Permission granted for use of the early logo from Continental Airlines, Inc.

Row 2 This logo is a registered trademark of Celgar Pulp Company.

Row 3 Permission granted from Northwest Hospital.

Logos, page 28 (listed left to right)

Row 1 The Star T design is a registered trademark of Texaco, Inc.
The Amtrak® corporate signature and the word Amtrak are registered service marks.
The CBS Eye design is a registered trademark of CBS Broadcasting, Inc.

Row 2 The Chevrolet bow tie emblem is a registered trademark of General Motors Corporation and is used by permission.
The Woolmark is used with permission from the Wool Bureau, Inc.

Row 3 The Pentastar logo is used with permission from the Chrysler Corporation.
The Checkerboard Square is a registered trademark of Ralston Purina Company.
The Hawaii Seafood logo is used with permission from Ocean Resources Branch, State of Hawaii Department of Business, Economic Development, and Tourism.

Row 4 Permission granted from Stanford Shopping Center.
Permission granted from Seville Properties.
Permission granted from Frost Bank.

Grateful acknowledgment is made to Linda Silvey and Loretta Taylor for the use of some ideas from *Paper and Scissors Polygons and More,* from Dale Seymour Publications.

Package graphic of Sour Cream 'N Chives Snack Crackers, page 83, used with permission from Safeway, Inc.

Honeycomb, page 84, used with permission from Monika Popowitz.

Bubbles photograph, page 85, by S. Schwartzenberg ©1988, The Exploratorium. http://www.exploratorium.edu.

For the following:
Symmetry Drawing E 123, page 92
Symmetry Drawing E 85, page 93
Symmetry Drawing E 75, page 158
Symmetry Drawing E 25, page 165
Symmetry Drawing E 67, page 174
Reptiles, page 40

Symmetry Drawing E 128, page 125
Symmetry Drawing E 105, page 155
Symmetry Drawing E 104, page 162
Symmetry Drawing E 99, page 168
Symmetry Drawing E 97, page 171
Alhambra Mosaic, page 124

M. C. Escher ©1998 Cordon Art B.V.-Baarn-Holland. All rights reserved.

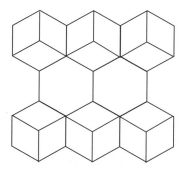

AUTHOR'S NOTE

Each July, I run a summer camp for middle school students. Two groups of 28 campers (plus volunteer teachers' aids) spend a full week exploring mathematical patterns and beach biology. The material in this book details the math content for the camp. The activities, however, are an ideal supplement to any math program. I recommend using them as a break from the daily math routine. You may want to give students something to look forward to by using the activities for "Friday-only" math classes.

Many of today's students are visual learners. In order to address their needs to learn through a variety of visual experiences, these activities incorporate the use of the overhead projector, hands-on manipulatives, computer software, and the Internet.

Although there is a lot of material, activities fly by because the students find them interesting (and fun!).

The activities consist of the following sections:

Materials   This is a list of the specific items you need for the activity. An overhead projector is required for all activities.

Vocabulary   This is a list of the terms used in the activity. Words are listed here only when they are being used for the first time in the book.

Activity Sequence   This section is the heart of the activity. It is here that I discuss all of the elements of the task. I have included questions (with answers) that you may discuss with your students. In some cases I have provided you with the results of my own teaching experiences.

Learning the Language   In this section, there is a definition for each of the terms listed in *Vocabulary*. If there is no vocabulary list in the activity, this section is omitted.

Student Challenge   This section is written to the student. You may use these as follow-ups or extensions of the activities.

The *Materials* section is especially important because you will need to gather the items prior to engaging your students in the activity. Some of the materials can be found in any mathematics classroom— rulers, scissors, crayons or colored pencils, pattern blocks, tape, overhead markers, and so on.

I have also listed the blackline masters needed for each activity. If the materials list says "transparency" after the master title, make a copy of it on clear acetate to use it as an overhead transparency for classroom discussion. (Please feel free to copy *all* blackline masters onto acetate for classroom application. I have only listed the ones that are essential to each activity.) If the materials list says "handout" after the master title, make a copy for each student and hand out the copies when told to do so.

All masters are numbered for easy reference. Several teachers' aids have referred to the blackline masters in this book as a visual treasury. The masters are presented in the sequence in which they are used in the camp, a sequence I find inherently logical.

In addition to the blackline masters, the following items are included in the envelope attached to the back cover of this book:

- A sample set of $3\frac{1}{2}$-inch-by-$2\frac{1}{4}$-inch plastic mirrors (they come in packs of four, but students will only need three mirrors for the activities in this book), two of which must be hinged (see instructions in Activity 8). I recommend you have enough mirrors available so your students may work in pairs. A trio of mirrors per student would be ideal. (Each student should, however, have his/her own copy of the required handouts.) Additional mirrors are available from Dale Seymour Publications.

- A sheet of tessellating shapes on translucent $8\frac{1}{2}$-inch-by-11-inch drawing paper. The blackline master appears in Activity 29 on page 175. Simply copy the master onto the drawing sheets as you would with ordinary paper. (See Appendix for information on paper stock.)

- A folded sheet of 11-inch-by-17-inch dot paper that can be used as a blackline master for the ink-printing exercise in Activity 28.

Some activities require special materials. In Activity 16, students need a bubble solution (see page 82 for details). In Activity 17, students make kaleidoscopes (see page 88). Students also need two rubber bands to make a prism kaleidoscope. Activities 27 and 28 require compressed sponge, self-adhesive foam rubber, and transparent stamp mounts, all of which are available from Dale Seymour Publications. Extra-large ($4\frac{1}{2}$-inch by $7\frac{1}{2}$-inch) stamp pads are also required for Activity 28. Activity 30 requires the demo version of *TesselMania!*® software (available through an Internet site), or the full versions of either *TesselMania!*® or *TesselMania!*® *Deluxe* software.

The *Appendix* (page 179) lists supplementary materials that are not essential to the activities, including posters and related Internet sites. I have created a homepage with links to these sites and I intend to update the links periodically.

What is Mathematics?

ACTIVITY SEQUENCE

Begin the lesson with the following question.

1. *In your own words, what is mathematics?*

    Most middle school students will equate mathematics with its computational tools ("things you can do with numbers") and have little, if any, understanding of the true nature of the subject. Many of today's mathematics educators choose to use the following definition:

    ### Mathematics is the study of patterns.

Show the transparency of the drawing of Pythagoras (Master 1A). The drawing of Pythagoras ("Pitagoras") is from a fifteenth-century woodcut. For Greeks such as Pythagoras, who lived more than 2,500 years ago, mathematics included music and astronomy. The computational tools of these ancient Greeks were few, yet they used mathematics to explore the world in which they lived. Mathematics was more than numbers and what you could do with them.

As mathematicians developed more and more analytical tools in ensuing centuries, mathematics education focused on drill-and-practice to the eventual virtual exclusion of any exploration of real-world phenomena. Math anxiety became a disease of epidemic proportions. Few would argue that the response would have been any different to a music curriculum that explored only scales, or a reading curriculum with spelling drills but no stories.

Today we are in the midst of a mathematical renaissance. By introducing calculators and computers into the curriculum, mathematics teachers can relegate their computational tools to a more appropriate reverence. Mathematics educators are endeavoring to rediscover their subject and its connections to the real world.

Give students the opportunity to write an answer to the following question. Then ask them to come up one at a time to record their answers on the chalkboard.

2. *Are there any objects with a pattern in this classroom? If yes, what are they?*

   Square or rectangular ceiling and floor tiles are a good place to start. Displaying pictures, posters, or actual examples of butterflies, honeycomb, snowflakes, flags, patchwork quilts, hubcaps, seashells, and so forth, will ensure a productive treasure hunt.

LEARNING THE LANGUAGE

Patterns are configurations of numbers, pictures, words, or any other things that have distinctive and consistent arrangements.

STUDENT CHALLENGE

Find real-world examples of objects that have a pattern. Be ready to share your examples and write a description of the patterns.

   Some of my students have produced lace doilies, crystals, an inlaid Islamic box, a pineapple, a chambered nautilus, a Ukrainian Easter egg, and even a wasp's nest! Students could share their examples in an old-fashioned "show and tell."

ACTIVITY

2 Introduction to Symmetry

MATERIALS

Master 2A, page 7
 (transparency)
Master 2B, page 8
 (transparency and
 handout)
Master 2C, page 9
 (transparency)
Rulers, I per pair
Overhead marker
Mirrors, I per pair

VOCABULARY

line of reflection
reflectional symmetry
vertical (reflectional)
 symmetry
horizontal (reflectional)
 symmetry
rotational symmetry

ACTIVITY SEQUENCE

Show the transparency of Master 2A. Discuss the following questions with your class.

1. What do these two drawings have in common?

Each can be cut precisely in half by a vertical line.

Ask a student to come up to the overhead to add the vertical line to the butterfly that divides the drawing into halves that exactly match. Have the student place a mirror upright and lengthwise along the line. Tell the student to look in the mirror to experience the *reflectional symmetry*. Since both the line and mirror are vertical, the butterfly has *vertical reflectional symmetry*. Invite another student to repeat the activity with the clown face. [If provided with a handout of Master 2A and sufficient mirrors to work in pairs, students can also do these activities at their desks as you demonstrate at the overhead.]

After placing the mirror, students will notice that the features on one side of the drawing have matching features on the other side. If you were to fold the acetate along the line, one half of both the butterfly and clown face would coincide with the other half. [You can demonstrate this with a duplicate of the transparency scored from upper to lower edge along the common vertical line with a sharp needle to permit folding.] Each reflected image looks identical to the design behind or "in" the mirror.

Show students that flipping the drawings about the common vertical line (demonstrate with transparency), would result in a pattern exactly like the original. Tell them the lines they have drawn are called *lines of reflection* or mirror lines.

It is important for students to understand there can be more than one line of reflection and the orientation is not always vertical.

Allow them to discover *horizontal reflectional symmetry* for themselves. Give each student a copy of Master 2B and give pairs of students a mirror and ruler. Show the transparency of Master 2B, then discuss the questions that follow.

2. *How many lines of reflection can you find? Add all the lines of reflection to the art and verify your additions with the mirror.*

The signs of the man and airplane have vertical reflectional symmetry. The sign of the telephone has horizontal reflectional symmetry. The designs in the middle have BOTH vertical and horizontal (reflectional) symmetry and the designs at the bottom lack any reflectional symmetry at all. Demonstrate the symmetries by adding the lines of reflection. Flip the transparency as required.

Master 2B provides an opportunity to introduce *rotational symmetry* in addition to reflectional symmetry. Tell students a figure has rotational symmetry if it appears the same after being turned less than 360° (a full turn). [The turn must be less than a full one; otherwise ALL figures would have rotational symmetry.] Several of the drawings in Master 2B will appear the same after a half turn.

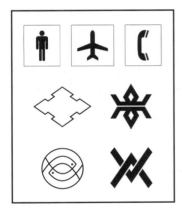 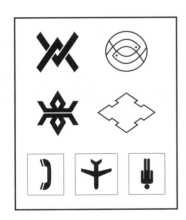

Rotate the transparency of Master 2B 180°. Students should maintain the handout in its original orientation for comparison, or switch your roles if preferred.

3. *Explain what you notice about the images.*

All but the signs look exactly the same. The ones that look the same are said to have rotational symmetry. The signs at the top of the page have reflectional symmetry, but no rotational symmetry. The designs at the bottom have rotational symmetry, but no reflectional symmetry. The designs in the middle have both kinds of symmetry.

You may wish to use a duplicate of the transparency as an overlay to demonstrate that a design with rotational symmetry will coincide with itself after an appropriate rotation.

LEARNING THE LANGUAGE

A *line of reflection* is a line that divides a figure or design into halves that are mirror images.

Reflectional symmetry is a characteristic of a figure or design that has at least one line of reflection.

Vertical (reflectional) symmetry is a characteristic of a figure or design that has a vertical line of reflection.

Horizontal (reflectional) symmetry is a characteristic of a figure or design that has a horizontal line of reflection.

Rotational symmetry is the characteristic of a figure or design that will coincide with (look the same as) itself after being turned less than 360° (a full turn).

Show the transparency of Master 2C and then have students respond to the following challenge.

STUDENT CHALLENGE

A magician placed the four cards pictured on a table. Blindfolded, he instructed someone from the audience to come up on stage to rotate one card 180°. The magician removed his blindfold. The cards as a group appeared unchanged. Can you figure out which card was turned? Write a response and justify your answer using one of the vocabulary words discussed during this activity.

> The answer is the eight of diamonds. It is the only card with rotational symmetry.

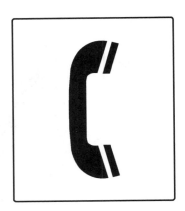

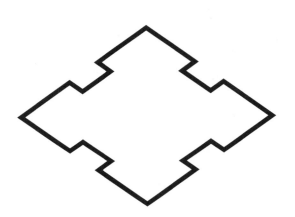

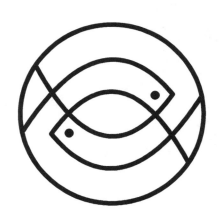

ACTIVITY 3

Symmetry in the Alphabet

Master 3A, page 12
    (transparency)
Master 3B, page 13
    (transparency)
Master 3C, page 14
    (transparency and
    handout)
Master 3D, page 15
    (transparency)
Mirrors, 1 per pair
Overhead marker

ACTIVITY SEQUENCE

Show the transparencies of Master 3A and Master 3B. The drawing of the word *mirror* comes from *Inversions,* a book by talented graphic artist, Scott Kim. Not only does the drawing read as the word *mirror,* but it also has reflectional or "mirror" symmetry. Scott is able to change m's into r's and other mathmagic.

Master 3B also comes from *Inversions* by Scott Kim. It reads *upside down* both right-side up and upside down. Demonstrate this rotational symmetry by turning the transparency 180°.

Now that students have seen a bit of symmetry in the alphabet, hand out mirrors and copies of Master 3C. Ask them to use the mirrors to answer the following questions. As students proceed, you may want to demonstrate the answers on the transparency by adding lines through the letters to show the vertical and horizontal reflection. Flip and turn the transparency as required to help students find the letters with rotational symmetry.

1. *Which letters have vertical reflectional symmetry?*

   A, H, I, M, O, T, U, V, W, X, Y

2. *Which letters have horizontal reflectional symmetry?*

   B, C, D, E, H, I, K, O, X

3. *Rotate the handout 180°. Which, if any, of the letters have rotational symmetry?*

   H, I, N, O, S, X, Z

4. *Which letters have both vertical and horizontal symmetry?*

   H, I, O, X

5. *Do the letters that have both vertical and horizontal symmetry also have rotational symmetry?*

   Yes, they all do.

6. *Which letters have rotational symmetry without reflectional symmetry?*

   N, S, Z

Students should learn that words can have symmetry. Display the transparency of Master 3D; then ask the following questions.

7. *What kind of symmetry does the word MATH have? What about the word CODE?*

   MATH has vertical symmetry. CODE has horizontal symmetry.

8. *What type of symmetry does SOS have?*

   SOS has rotational symmetry, whether written vertically or horizontally (demonstrate by rotating the transparency 180°).

9. *What type of symmetry does IXOHOXI have?*

   It has vertical reflectional symmetry, horizontal reflectional symmetry, and rotational symmetry.

STUDENT CHALLENGE

Create your own alphabet symmetry examples. Use appropriate letters and describe the type or types of symmetry exhibited by your creation.

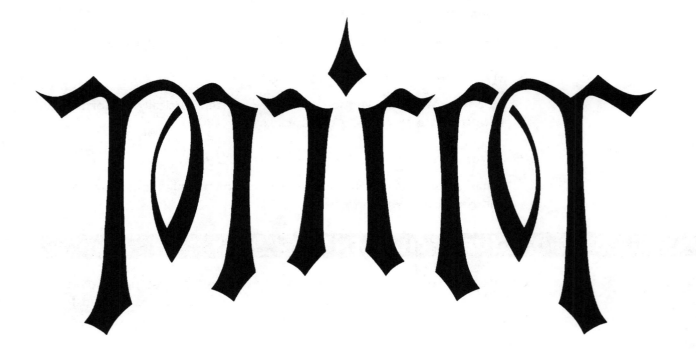

upside down

A B C D E
F G H I J K
L M N O P
Q R S T U
V W X Y Z

ACTIVITY 3 MASTER 3C Investigating Patterns/**Symmetry and Tessellations**

M   CODE
A
T   S
H   SOS
                 S

IXOHOXI

ACTIVITY 4

Symmetry in Flags and First Nations Art

MATERIALS

Master 4A, page 19
(transparency and
handout)
Master 4B, page 20
(transparency)
Master 4C, page 21
(transparency)
Mirrors, 1 per pair
Clear acetate, 1 per student
(optional)
Colored markers (optional)
Pattern blocks, 1 set per
group (optional)

VOCABULARY

angle of rotation
anti-symmetry

ACTIVITY SEQUENCE

Show the transparency of Master 4A. Give each pair of students a mirror and a copy of Master 4A. Discuss the following questions with the class, while demonstrating by flipping or turning the transparency as required.

1. *Many flags are symmetrical. Do any of these flags have vertical reflectional symmetry? Verify with the mirror.*

   The flags of Canada, Israel, and Switzerland have vertical reflectional symmetry.

2. *Do any flags have horizontal reflectional symmetry?*

   The flags of Norway, Israel, and Switzerland have horizontal reflectional symmetry. Incidentally, the star on the flag of North Korea breaks the otherwise horizontal reflectional symmetry.

3. *Do any flags have rotational symmetry?*

   The flags of Israel, Trinidad, and Switzerland have rotational symmetry.

4. *Do any flags have rotational symmetry, but no reflectional symmetry?*

   Trinidad's flag has rotational symmetry, but no reflectional symmetry.

5. *The flag of Switzerland has two additional lines of reflection. Can you find them with the mirror? Describe their locations.*

   One extends diagonally from the top-left to the bottom-right corner, and the other from the bottom-left to the top-right corner.

Explain to students that the Swiss flag looks exactly the same if you rotate it 180°, as do the flags of Israel (reflectional symmetry) and Trinidad (no reflectional symmetry). However the Swiss flag will look exactly the same if you rotate it just 90°. Indeed it will look the same if you rotate it 90°, 180°, 270°, or 360°. The smallest of these angles (*angle of rotation*) is 90°. The angle of rotation for the flags of Israel and Trinidad is 180°.

Show students the transparency of Master 4B. Ask them to notice that the flag of South Korea exhibits *anti-symmetry,* or symmetry with a property reversal. The circular symbol in the center, called the "yin-yang," represents opposing forces in the universe. It is anti-symmetrical because of the white/black interchange. The various markings (Pa-kua symbols) are opposites of one another diagonally. Broken lines are unbroken and vice versa. These symbols represent numbers in the binary (base two) number system. The first eight symbols and their meanings appear at the bottom of the graphic.

Have students interpret the symbols on the South Korean flag. Heaven at the upper left is opposite Earth at the lower right, and water at the upper right is opposite fire at the lower left. The sum of the numbers that correspond to each pair of opposite numbers is 7.

Show students the transparency of Master 4C.

This graphic is called "Our Beginnings." It was created by Art Thompson, a Nuu-chah-nulth Salish artist. Salish Native American art reflects the style of many Northwest Coastal First Nations artwork. Unlike this example, most of the art is "balanced" but unsymmetrical. Although identical mirrored components will appear on each side of a totem pole or traditionally circular print, there will usually be distinct corresponding components as well.

Point out that Art Thompson's piece has vertical reflectional symmetry. The design depicts the Ditidaht creation story. In the story, Raven offers his knowledge of the world to the people for safe-keeping. Over the course of generations, this knowledge is passed on, and the spirit of transformation begins to unfold.

LEARNING THE LANGUAGE

The *angle of rotation* of a design with rotational symmetry is the smallest angle through which it can be turned so as to coincide with (look the same as) itself.

Anti-symmetry is symmetry with property reversal. The object or drawing would be symmetrical if not for specific images that are opposites instead of identical.

STUDENT CHALLENGE

Patchwork quilt blocks—the design element that repeats throughout a quilt top—often exhibit reflectional and/or rotational symmetry. Reconstruct one or more classic symmetrical quilt block patterns on grid paper.

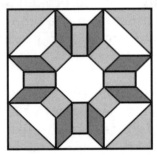
Carpenter's Wheel

Prosperity Block

You may want to give the students clear acetate and colored markers to transfer their designs and use them as window decorations. If you have pattern blocks, allow students to use them to recreate a symmetrical quilt block, or have them create a quilt block of their own design.

Canada

North Korea

Norway

Israel

Trinidad

Switzerland

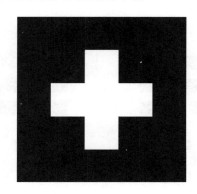

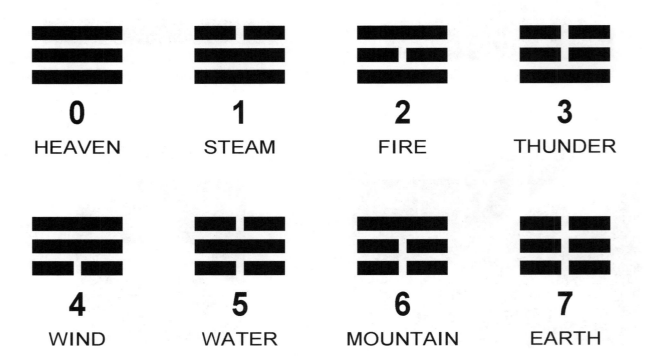

0

HEAVEN

1

STEAM

2

FIRE

3

THUNDER

4

WIND

5

WATER

6

MOUNTAIN

7

EARTH

More on Reflectional and Rotational Symmetry

ACTIVITY SEQUENCE

Show the transparency of Master 5A. Explain that each of the designs on the left side of the page has both reflectional and rotational symmetry. Each of the designs on the right side of the page has rotational symmetry without reflectional symmetry. As a class, add all lines of symmetry to the designs on the left. Ask students the following questions.

1. *How many lines of symmetry do the various designs have?*

   Top left—3, middle left—5, bottom left—6. You may need to demonstrate by drawing the lines on the transparency.

2. *Both of the designs in each of the three rows have the same angle of rotation. The simpler designs on the left should help you analyze the more complex ones on the right. In how many different ways can each design look exactly like the original (coincide with itself)?*

   Turn the transparency, remembering to count each original position exactly once. Demonstrate coincidence using a duplicate of the transparency as an overlay. The number of identical points should provide a clue. The top row has 3, the middle row has 5, and the bottom row has 6.

Tell students this is called the *order of rotation*. The first design has an order of rotation equal to 3. The point about which each design turns is called the *center of rotation*.

3. *If the angle of rotation is 180°, the order is 2. If the angle of rotation is 90°, the order is 4. Describe the relationship between the order and the angle of rotation.*

   The product of the order and the angle of rotation is always 360°. Equivalently, the angle of rotation is 360° divided by the order of rotation.

MATERIALS

Master 5A, page 25
   (transparency,
   two copies)
Master 5B, page 26
   (transparency)
Master 5C, page 27
   (transparency)
Master 5D, page 28 (handout)
Rulers, 1 per student
Mirrors, 1 per pair
Overhead marker

VOCABULARY

order of rotation
center of rotation

4. Determine the angles of rotation for the six designs.

The first row is $\frac{360°}{3} = 120°$. The second row is $\frac{360°}{5} = 72°$. The third row is $\frac{360°}{6} = 60°$.

5. For designs with reflectional and rotational symmetry, describe the relationship between the number of lines of reflection and the order of rotation.

They are equal.

Tell students that decoration and shading can reduce the order of a design. This is evident in the Pennsylvania hex signs on Master 5B. Show this transparency to students. Without the flowers, the sign at the top would have order 8. With the flowers, the order is reduced to 4. Adding shading, as in the sign at the bottom, reduces the order to 2. See the Appendix for sources of additional hex signs to explore.

Show the transparency of Master 5C. Commercial trademarks or logos and representative symbols or emblems are often symmetrical. Discuss the following questions with your class.

6. Can you identify any of the symbols on Master 5C? Comment on the symmetry in each.

In the top row, the Volkswagen symbol on the left has vertical reflectional symmetry. The former logo of Continental Airlines on the right has horizontal reflectional symmetry, as does the artistic logo of the Celgar Pulp Company in the middle. In the last row, the logo on the left belongs to Northwest Hospital in Seattle. Notice the hospital cross in the center and the people surrounding it. The graphic has four lines of reflectional symmetry and rotational symmetry of order 4. The familiar recycling symbol next to it has rotational symmetry of order 3.

As a follow-up exercise, give each pair of students a mirror and a copy of Master 5D.

7. Add the lines of reflection to the graphics that have reflectional symmetry and determine the appropriate order for those that have rotational symmetry. Try to identify familiar logos.

Row one has the logos of Texaco (vertical reflectional), Amtrak (horizontal reflectional), and CBS (vertical and horizontal reflectional, rotational of order 2). Row two has the radioactive symbol (three lines of reflection, rotational of order 3), the logo of Chevrolet (rotational of

order 2), and the common Woolmark (rotational of order 3). The
third row has the logos of Chrysler (five lines of reflection, rotational
of order 5), Ralston Purina (four lines of reflection, rotational
of order 4), and Hawaii Seafood (rotational of order 5). Row four shows
the logos of the Stanford Shopping Center (rotational of order 4),
Seville Properties (rotational of order 8), and Frost Bank (rotational
of order 19).

LEARNING THE LANGUAGE

The *order of rotation* of a design with rotational symmetry is the
number of distinct times it can coincide with (look the same as)
itself after successive rotations through the angle of rotation.

The *center of rotation* of a design with rotational symmetry is the
point about which the design turns to achieve coincidence.

STUDENT CHALLENGE

Are there any other logos that you could think of that are
symmetrical? If yes, draw one and describe the type of symmetry.
If no, find one in the newspaper or the yellow pages and describe
its symmetry.

ACTIVITY 5 MASTER 5B   Investigating Patterns/**Symmetry and Tessellations**

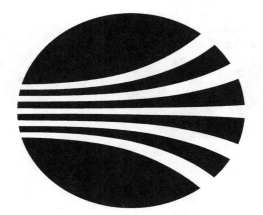

ACTIVITY 5 MASTER 5D   Investigating Patterns/**Symmetry and Tessellations**

Symmetrical Strip Patterns

ACTIVITY SEQUENCE

MATERIALS

Master 6A, page 32
     (transparency, 2 copies)
Master 6B, page 33
     (transparency, 2 copies
     if needed)
Scissors

VOCABULARY

translation
glide reflection
translational symmetry
glide-reflectional symmetry
strip pattern

In advance, prepare two transparencies of Master 6A (animal tracks). Make three overlays by cutting apart the three columns of patterns on one of the transparencies.

Show your class the uncut transparency of Master 6A. Ask students to imagine that the three patterns extend upward and downward indefinitely.

Use the cutouts as overlays. Demonstrate how sliding the overlays up or down extends the pattern. Each print or pair of prints represent a basic design element. Sliding the design element in a fixed direction creates the pattern. Each pattern is said to have *translational symmetry*.

Discuss the following questions with your class.

1. *What can we deduce about the animal making each of the tracks? What type of symmetry does each set of tracks exhibit?*

   The tracks on the left were made by a biped hopping on one foot. The pattern has translational symmetry. The tracks in the middle were made by a biped hopping on both feet. Since the animal's feet are always directly opposite one another, the pattern has reflectional as well as translational symmetry. The vertical line shown is the line of reflection. The tracks on the right were made by a biped with a gait similar to that of a human being. The pattern has translational symmetry, but lacks reflectional symmetry.

Tell students you can, however, make the pattern on the right coincide with itself by reflecting it in the line shown, and then sliding it in a direction parallel to this line (demonstrate with the third piece of overlay). The sequence, reflecting and sliding, can be reversed. The motion is called *glide reflection*. Such a pattern is said to have *glide-reflectional symmetry*.

Incidentally the pattern in the middle also has glide-reflectional symmetry, as well as translational and reflectional symmetry (demonstrate with the second piece of overlay). Any pattern with both translational and reflectional symmetry will have glide-reflectional symmetry. On the other hand, the pattern on the right has translational and glide-reflectional symmetry without reflectional symmetry.

Show the transparency of Master 6B then discuss the following question with your class.

2. *Consider each pattern as extending to the left and right indefinitely. Each is referred to as a strip pattern, or frieze. What kind of symmetry does each strip pattern exhibit?*

   The first has translational symmetry. The second has translational symmetry and vertical reflectional symmetry. The third has translational symmetry, horizontal reflectional symmetry, as well as glide-reflectional symmetry. The fourth has translational symmetry and glide-reflectional symmetry. (If needed, demonstrate the symmetries with a duplicate of the transparency as an overlay.)

There are seven unique combinations of symmetries that can be found in strip patterns. For more information on these combinations, see the Appendix.

LEARNING THE LANGUAGE

A *translation* is a motion that slides a figure or design in a fixed direction.

A *glide reflection* is a motion that reflects a figure or design in a straight line, then slides it in a direction parallel to that line.

Translational symmetry is the characteristic of a design that is able to coincide with (look the same as) itself after sliding in a fixed direction.

Glide-reflectional symmetry is the characteristic of a design that is able to coincide with (look the same as) itself after reflection in a straight line, followed by sliding in a direction parallel to that line.

A *strip pattern* or *frieze* is a design that exhibits translational and/or glide-reflectional symmetry in just one fixed direction.

Trace the butterfly pattern below on a 1-foot-long rectangular strip of $2\frac{1}{4}$-inch-wide adding machine tape, matching the left side of the pattern to the left edge of the strip. Fold the paper backwards along the right side of the tracing, then fold it back and forth like a fan, keeping all edges even as you fold until the length of the strip is exhausted.

Cut away the shaded areas through all layers, then open up your paper frieze. If the pattern could be extended forever, what kind of symmetry would it have? (Answer: translational and vertical reflectional symmetry.)

Repeat for either the teddy bear or the frog (use a hole punch for the eye). If you wish, create your own pattern, making sure the left and right sides have fold areas that don't get cut.

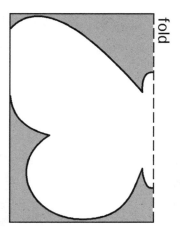
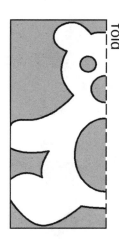
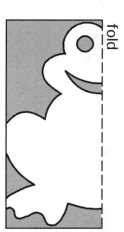

ACTIVITY 6 MASTER 6A   Investigating Patterns/**Symmetry and Tessellations**

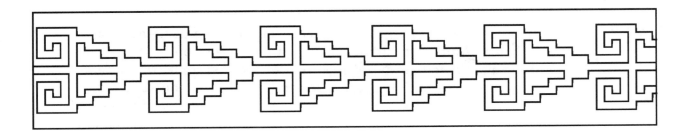

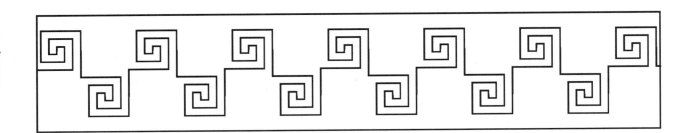

ACTIVITY 7

Introduction to Polygons

ACTIVITY SEQUENCE

MATERIALS

Master 7A, page 38 (handout)
Master 7B, page 39
    (transparency)
Master 7C, page 40
    (transparency)
Master 7D, page 41
    (transparency and
    handout)
Examples of triangular support
    (pictures of Eiffel Tower,
    tripods, and so on)
Straws (about 20)
String
Overhead marker

VOCABULARY

vertex
polygon
scalene triangle
isosceles triangle
equilateral triangle
quadrilateral
rhombus
parallelogram

In advance, prepare a set of straw polygons by threading straws on string, then tying each assembly with a secure knot. Include a scalene triangle, isosceles triangle, equilateral triangle, rhombus (four equal lengths), and parallelogram with two different lengths.

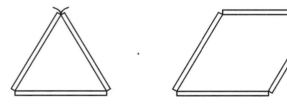

Discuss with students the popular children's puzzle consisting of a set of numbered dots that are joined in sequence to make a picture. If the first point is joined to the last point to make a closed figure, the result is referred to as a polygon. The term *polygon* is derived from a Greek word meaning "many angled." The numbered points are the vertices (singular: *vertex*) of the polygon. The line segments joining the points are the polygon's sides.

Give each student a copy of Master 7A. Have students complete the connect-the-dots puzzle. (Remind them to close the figure by joining dot 50 to dot 1.)

After they have finished, show them the transparency of Master 7B (the completed puzzle). The polygon in the solution resembles a lizard or reptile.

The creature is an adaptation of one that appears in the lithograph "Reptiles" by the Dutch graphic artist M. C. Escher

(1898–1972). It is one of several identical lizards assembled in a jigsaw puzzle configuration in Escher's own sketchbook (display the transparency of Master 7C). It depicts a succession of lizards crawling to life, climaxing in a triumphant snort, then returning to their two-dimensional world. Students will discover the mathematical basis of Escher's lizard in Activity 29.

Inform students that the simplest polygon is the familiar triangle, meaning "three angles." (Show students all three straw triangles.) Point out that whether it has no equal sides (*scalene*), two equal sides (*isosceles*), or three equal sides (*equilateral,* meaning "equal side"), a triangle made out of straws is rigid.

Show students either of the straw *quadrilaterals*. Straw quadrilaterals, meaning "four sides," are not rigid (demonstrate). A triangle is the only polygon made up of jointed strips that is intrinsically rigid. In other words, a triangle whose side lengths are specified can have one, and only one, possible shape. If the side lengths are fixed, so are the angle measures, and vice versa.

This simple geometrical fact finds practical applications in reinforcing structures, and in the use of guy ropes and props. Discuss with students how triangles appear in roof construction and in the shape of lawn chairs and stepladders. It explains the diagonal bar on farm gates and the cross-girdering on rigid structures, including cranes, girder bridges, and even immense structures such as the Eiffel Tower. You may want to show your students pictures of the Eiffel Tower and other examples of triangular rigid supports.

Have students answer the following questions as a class.

Hold up the straw *rhombus* so the straws are coplanar, and the quadrilateral has no right angles.

1. *What do you call a quadrilateral with four equal sides?*

   A rhombus—<u>not</u> a square!

Hold up the straw *parallelogram* so the straws are coplanar, and each angle is a right angle.

2. *What do you call a quadrilateral with four equal angles?*

   A rectangle—<u>not</u> a square!

Hold onto the upper and lower straws of the rectangle, then gradually move them in opposite directions while maintaining a coplanar configuration. Notice that as you move the straws, the opposite sides of the quadrilateral stay parallel. A quadrilateral with two sets of parallel sides is called a *parallelogram*. Opposite sides of a parallelogram are always equal as well as parallel. Return the angles of the quadrilateral to right angles. A rectangle is both a quadrilateral and a parallelogram with four equal (right) angles.

Hold up the straw rhombus so the straws are coplanar, and the angles of the quadrilateral are right angles.

3. *What do you call a quadrilateral with four equal sides AND four equal angles?*

   a square

A square is a rhombus with four equal angles and a rectangle with four equal sides. It is both a quadrilateral and a parallelogram as well.

Summarize the findings by giving students a copy of Master 7D. Ask them to write in the names of each figure. Fill in the blanks on the overhead transparency of Master 7D. All four figures are parallelograms as well.

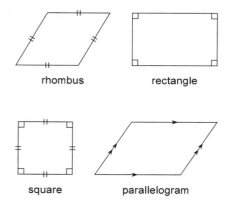

rhombus   rectangle

square   parallelogram

LEARNING THE LANGUAGE

The *vertex* of a polygon is the common endpoint of two of its sides.

A *polygon* is a closed geometrical figure with straight line segments as sides.

A *scalene triangle* is a triangle with three sides of unequal length.

An *isosceles triangle* is a triangle with two sides of equal length.

An *equilateral triangle* is a triangle with three sides of equal length.

A *quadrilateral* is a polygon with four sides and four angles.

A *rhombus* is a quadrilateral with four sides of equal length.

A *parallelogram* is a quadrilateral with parallel opposite sides.

STUDENT CHALLENGE

Create your own connect-the-dots puzzle. Record the answer on a separate piece of paper, then exchange your puzzle with a partner and try to solve each other's creations.

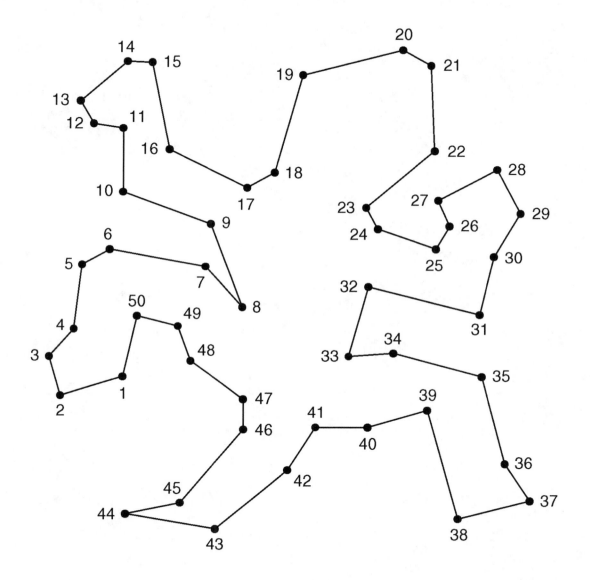

ACTIVITY 7 MASTER 7C Investigating Patterns/**Symmetry and Tessellations**

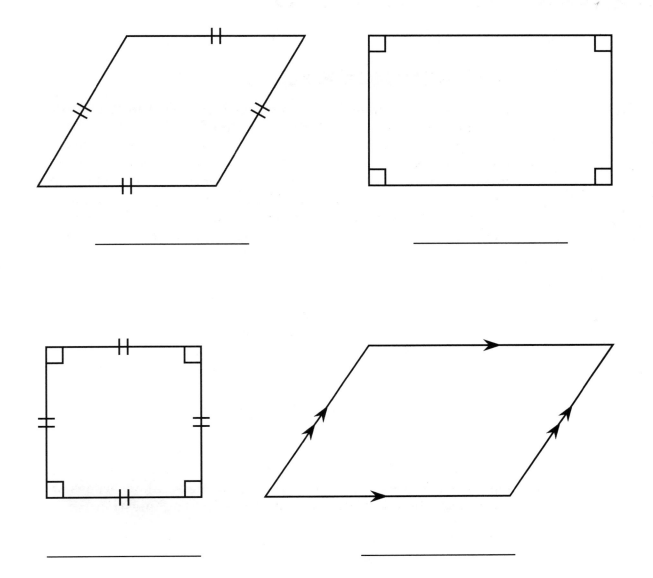

ACTIVITY

8

A Regular Polygon Kaleidoscope

MATERIALS

Master 8A, page 45 (handout)
Master 8B, page 46
    (transparency and
    handout)
Master 8C, page 47 (handout)
Master 8D, page 48
    (transparency)
Mirrors, 2 per pair of students
$\frac{3}{4}$-inch cloth tape cut into $2\frac{1}{4}$-inch
    long pieces, 1 piece per
    pair of students
Protractors, 1 per student
    (optional)

VOCABULARY

regular polygon
pentagon
hexagon
mirror angle

ACTIVITY SEQUENCE

Prepare a set of hinged mirrors for each pair of students. To do this, place two mirrors face to face with edges flush. Use $\frac{3}{4}$-inch cloth tape to attach the pair together, forming a hinge on either short side (see below). Lay the assembly flat, reflective surfaces down, and push the two hinged edges together. This will secure the hinge for precision results. [Students will repeat this last step each time they begin a new hinged mirror activity.] This assembly is called a "hinged-mirror kaleidoscope" throughout this book.

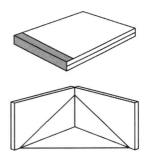

Give each pair of students a hinged-mirror kaleidoscope and a copy of Master 8A. Have students stand their kaleidoscopes on the broken lines and then discuss the following questions.

1. *When you look straight into the kaleidoscope (not down at it), what is the figure that you see?*

    an equilateral triangle (see above)

Tell students that an equilateral triangle is a *regular polygon*. A polygon is regular if all of its sides are the same length and all its angles are the same measure. If all three sides of a triangle are equal, then its angles must also be equal, and vice versa. The triangle is the only geometric figure, however, for which this is true (see Activity 7). The properties of being equilateral and equiangular in a geometric figure are independent. A polygon must possess both to be called regular.

Have students move the mirrors toward one another until they see a square. Have them move the mirrors closer again until they see, in turn, a regular *pentagon,* a regular *hexagon,* a regular heptagon, a regular octagon, a regular nonagon, and a regular decagon. Ask the following question.

2. *Why do you think these shapes are called by these names?*

The names are taken from their number of sides. The figures were studied extensively by the ancient Greeks, and the names they gave them have been used ever since.

Give students copies of Master 8B. Have students fill in the correct names for each of the polygons pictured. When they finish, fill in the answers as a class on the transparency of Master 8B.

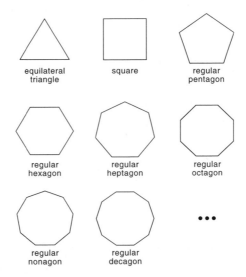

equilateral triangle	square	regular pentagon
regular hexagon	regular heptagon	regular octagon
regular nonagon	regular decagon	•••

Explain that the set of regular polygons, like the set of counting numbers, goes on indefinitely. The greater the number of sides, the more symmetrical it is, and the more nearly circular in shape. A regular polygon with, say, 100 sides (a regular 100-gon) would look so much like a circle that it might very well be mistaken for one.

If you draw line segments from the center (of rotation) of a regular polygon to each of its vertices, you divide the polygon into identical (congruent) isosceles triangles. The apex angle of any of these triangles (the angle at the center) is a *mirror angle* of the polygon. This is the angle between the mirrors when the corresponding regular polygon is formed by the hinged-mirror kaleidoscope.

Hand out copies of Master 8C. If you wish, have students use protractors to fill in the mirror angles for each of these shapes.

The mirror angle of an equilateral triangle has a measure of 120°, the mirror angle of a square has 90°, the mirror angle of a regular pentagon has 72°, and the mirror angle of a regular hexagon has 60°. These measures come from the fact that the number of degrees surrounding a point is always 360. Thus $120° = \frac{360°}{3}$, $90° = \frac{360°}{4}$, and so forth. In general, the mirror angle of a regular polygon having n sides has a measure of $\frac{360°}{n}$.

Show the transparency of Master 8D. Explain to students that they will find examples of equilateral triangles and squares everywhere. Many flowers and starfish are pentagonal. Snowflakes are shaped like regular hexagons, and stop signs are shaped like regular octagons.

LEARNING THE LANGUAGE

A *regular polygon* is a polygon with sides of equal length and angles of equal measure.

A *pentagon* is a polygon with five angles and five sides.

A *hexagon* is a polygon with six angles and six sides.

A *mirror angle* of a regular polygon is any of the identical (congruent) angles formed by joining its center (of rotation) to two adjacent vertices of the polygon.

STUDENT CHALLENGE

How many pentagonal objects can you find in the classroom? How about hexagonal objects? Make a list of the items you find.

The pentagon building in Washington, DC is in the shape of a regular pentagon. Find out why this is so.

> There are 5 branches of the American military (Army, Navy, Marines, Air Force, and Coast Guard).

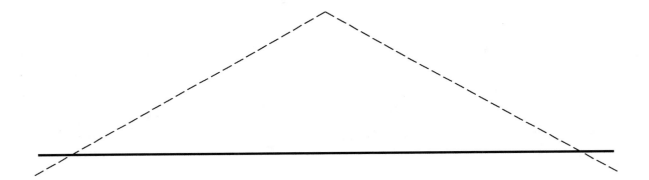

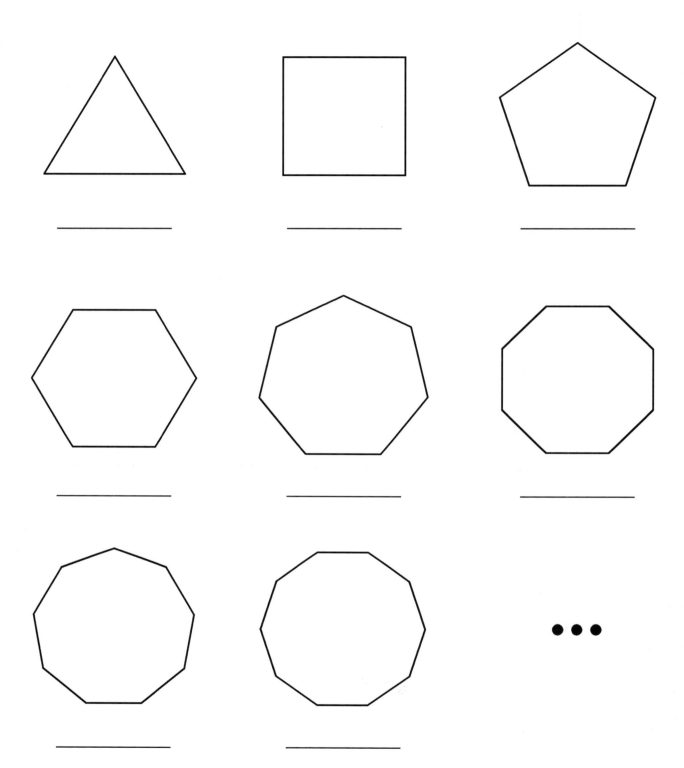

ACTIVITY 8 MASTER 8B Investigating Patterns/**Symmetry and Tessellations**

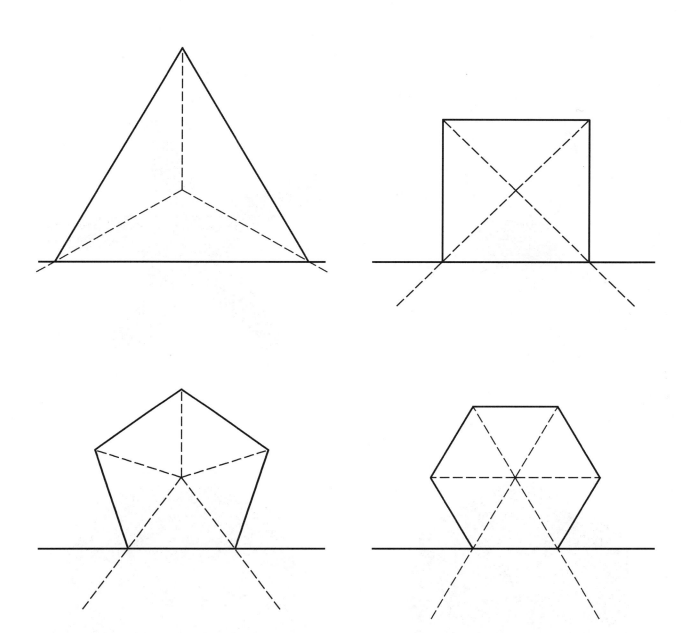

$$\textbf{measure of mirror angle} = \frac{360°}{n}$$

ACTIVITY 8 MASTER 8D   Investigating Patterns/**Symmetry and Tessellations**

Symmetry and Regular Polygons

ACTIVITY SEQUENCE

MATERIALS

Master 8B, page 46 (handout
    from Activity 8)
Master 9A, page 52 (handout)
Master 9B, page 53 (handout)
Master 9C, page 54 (handout)
Master 9D, page 55 (handout)
Hinged-mirror kaleidoscopes,
    I per pair
Pattern blocks (optional)

Regular polygons look more like a circle as the number of their sides increases because of their symmetries. Students should use their copies of Master 8B to answer the following question.

1. *Add all lines of reflection to each regular polygon. How many lines of reflection does each polygon have?*

   The equilateral triangle has three lines of reflection. The square has four. The regular pentagon has five, and so on. Help students see that if a regular polygon has *n* sides, it will have *n* lines of reflection, as well as rotational symmetry of order *n*.

Now hand out copies of Master 9A and have students answer the following question.

2. *Into how many identical (congruent) triangles do the lines of reflection divide the regular pentagon?*

   The pentagon is divided into ten identical triangles.

Next, ask students to place their hinged-mirror kaleidoscopes so they stand on any two adjacent lines of reflection with the hinges at the center of the polygon. (It will form the complete regular pentagon.) Ask students the following questions.

3. *What fraction of the pentagon lies between the mirrors?*
   $\frac{1}{10}$

4. *How large is the angle between the mirrors?*
   $\frac{360°}{10} = 36°$

Instruct students to keep one mirror fixed and pivot the other mirror around the hinge until it lands on the next line of reflection. They will see the regular pentagon again. This time there will be $\frac{2}{10} = \frac{1}{5}$ of the pentagon between the mirrors. Continue in this manner. They can generate the complete pentagon with $\frac{3}{10}$; $\frac{4}{10} = \frac{2}{5}$; $\frac{5}{10} = \frac{1}{2}$; $\frac{6}{10} = \frac{3}{5}$; and so forth, of the polygon between the mirrors.

Hand out copies of Master 9B. Have students place their hinged-mirror kaleidoscopes on two adjacent lines of reflection; then ask the following questions.

5. *Into how many identical (congruent) triangles do the lines of reflection divide the regular hexagon?*

   The hexagon is divided into 12 identical triangles.

6. *What fraction of the hexagon lies between the mirrors?*

   $\frac{1}{12}$

7. *How large is the angle between the mirrors?*

   The angle between any two adjacent lines of reflection is $\frac{360°}{12} = 30°$.

8. *You can see a kaleidoscope image of a complete hexagon with what portion of the polygon between the mirrors?*

   The complete hexagon can be generated with $\frac{1}{12}$; $\frac{2}{12} = \frac{1}{6}$; $\frac{3}{12} = \frac{1}{4}$; and so forth of the polygon.

Hand out copies of Master 9C. Have students position their hinged-mirror kaleidoscopes as before and then discuss the following questions.

9. *Into how many identical regions do the lines of reflection divide the Swiss flag?*

   The flag is divided into eight identical regions.

10. *What fraction of the flag lies between the mirrors?*

    $\frac{1}{8}$

11. *How large is the angle between the mirrors?*

    The angle between any two adjacent lines of reflection is $\frac{360°}{8} = 45°$.

12. *What is the smallest portion of the Swiss flag that can generate the entire flag in the hinged-mirror kaleidoscope?*

   One-eighth of the flag.

Hand out copies of Master 9D. Have students position their hinged-mirror kaleidoscopes on the top and then bottom figures of Master 9D to answer the following questions.

13. *The top figure will generate a five-pointed star, and the one at the bottom will form a ring of five paper dolls. How large is the angle that lies between the mirrors?*

   $\frac{360°}{5} = 72°$

14. *Both figures have vertical reflectional symmetry. If you pivot one of the mirrors closer to the other until it is vertical on the page, you will find that half of each region will also create the kaleidoscope images. How large is the angle between the mirrors now?*

   $\frac{360°}{10} = \frac{72°}{2} = 36°$

STUDENT CHALLENGE

Make your own "mirror" design using pattern blocks or cutouts of shapes placed between the hinged mirrors. Copy the complete design onto paper and add all lines of reflection. Describe some of the characteristics of your design and its symmetrical features.

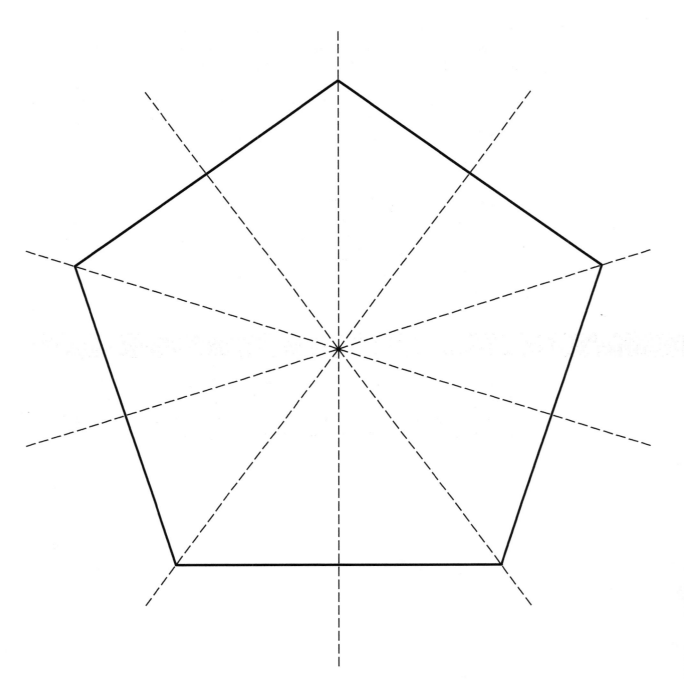

ACTIVITY 9 MASTER 9A   Investigating Patterns/**Symmetry and Tessellations**

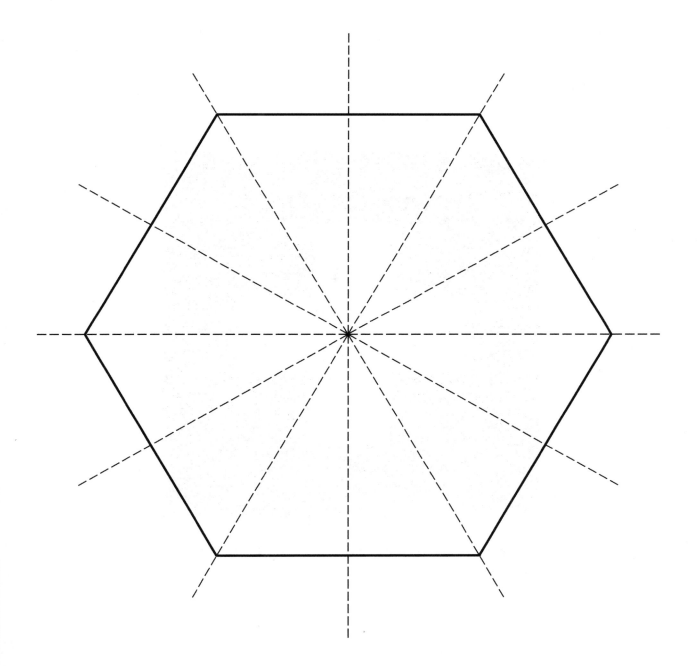

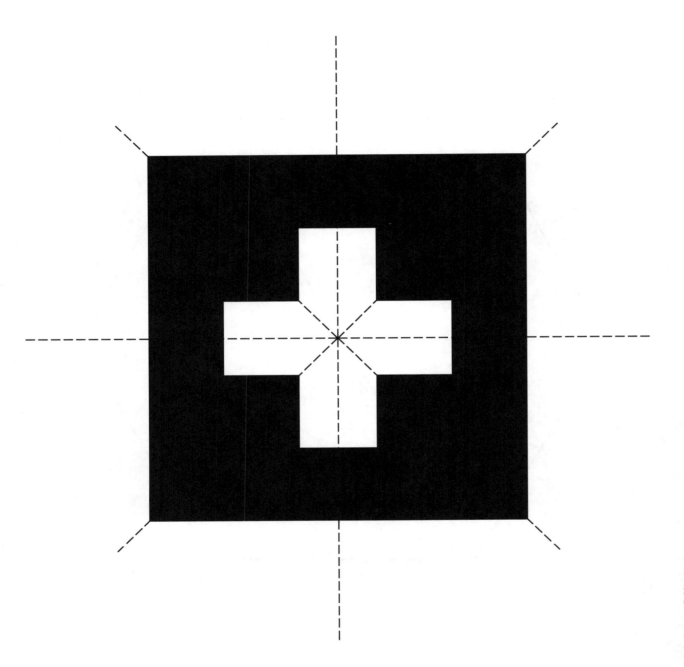

ACTIVITY 9 MASTER 9C Investigating Patterns/**Symmetry and Tessellations**

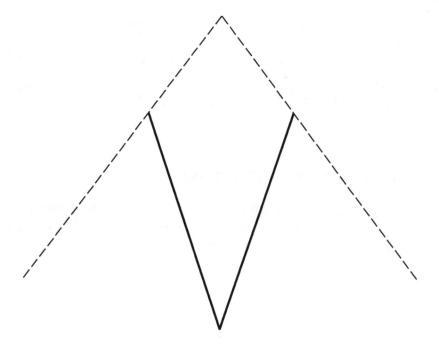

Symmetry by Paper Folding and Cutting

ACTIVITY SEQUENCE

MATERIALS

Master 10A, page 58
    (transparency)
Master 10B, page 59
    (transparency)
Scissors, 1 per student
Blank sheets of lightweight
    paper, several per student

Have students make symmetrical hearts by folding paper in half and cutting half the heart against the fold. If students need help, demonstrate with paper and scissors.

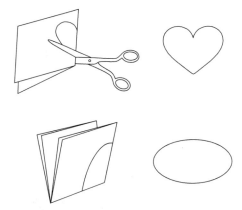

Tell students that if they fold two perpendicular creases, keep the paper folded, and then cut a curve, their cutout will have two perpendicular lines of reflection.

Incidentally, this is a simple way to cut a hole in a piece of paper or plastic—particularly with blunt scissors! Simply fold the paper or plastic in four, then snip off the corner where the folds meet.

Show students the transparency of Master 10A. It gives instructions for making a square by folding its four lines of reflection, and then cutting the paper. The instructions are self-explanatory. Lightweight pinfeed computer paper works best for this and the next two activities. Regular bond is too thick, newspaper is too dirty, and origami paper is simply too expensive!

Have students make a square by following the steps on Master 10A. After they have obtained △GFJ, tell students <u>not</u> to unfold the triangle. Point out that GF and GJ are the folded sides of the triangle, as indicated by the broken lines. They represent two adjacent lines of reflection of the square. Unfold a demonstration sample to reinforce this idea if required. Ask students the following question.

1. *How large is the angle between these two lines of reflection?*
$$\frac{360°}{8} = \frac{90°}{2} = 45°$$

Show the transparency of Master 10B and then instruct students to work on the following challenge.

STUDENT CHALLENGE

Cut out shapes (represented by the shaded regions) from the two folded edges. You will obtain a decorative square when you unfold the paper (the bottom diagram). Be careful not to cut across from one folded edge to the other. The drawing is merely an illustration. Please create your own.

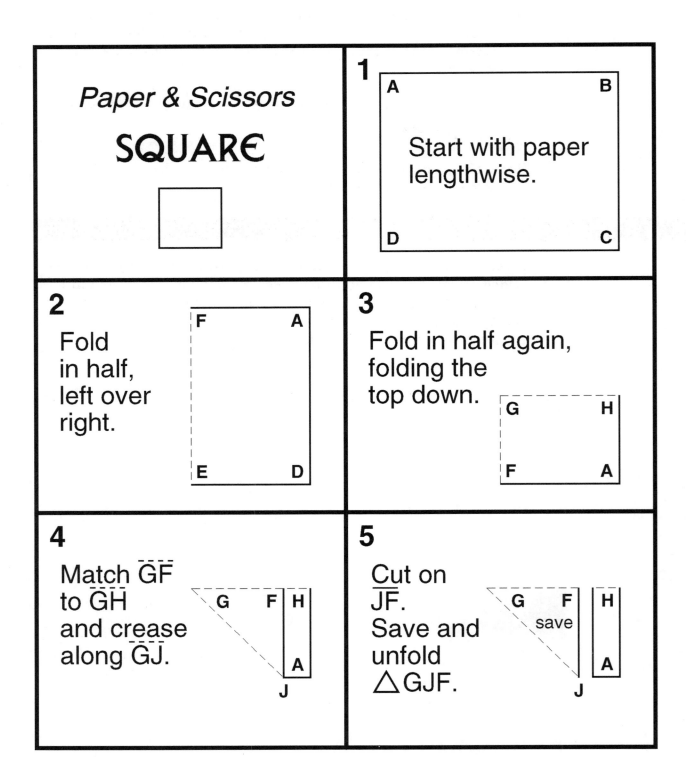

Paper & Scissors

SQUARE

1 Start with paper lengthwise.

A · B · D · C

2 Fold in half, left over right.

F · A · E · D

3 Fold in half again, folding the top down.

G · H · F · A

4 Match $\overline{GF}$ to $\overline{GH}$ and crease along $\overline{GJ}$.

G · F · H · A · J

5 Cut on $\overline{JF}$. Save and unfold △GJF.

G · F · save · H · A · J

© Dale Seymour Publications®

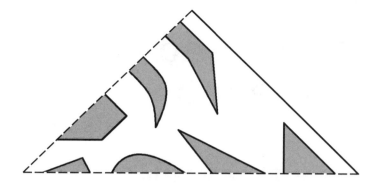

Investigating Patterns/**Symmetry and Tessellations** ACTIVITY 10 MASTER 10B

Paper and Scissors Pentagon

ACTIVITY SEQUENCE

MATERIALS

Master 9A, page 52 (handout
        from Activity 9)
Master 11A, page 62
        (transparency)
Master 11B, page 63
        (transparency)
Scissors, 1 per student
Blank sheets of lightweight
        paper, several per student

Have students look back at the drawing of a regular pentagon with its five lines of reflection from Activity 9 (Master 9A). Mention that we proved earlier that the angle between two adjacent lines of reflection of a regular pentagon is $\frac{360}{10}^{\circ} = 36^{\circ}$. To create a paper and scissors regular pentagon, the angle between the folded sides of the eventual triangle must be 36°.

Show students the transparency of Master 11A. Have them follow the directions to create a regular pentagon using a sheet of lightweight paper and scissors. The instructions for folding and cutting the regular pentagon are similar to those just used to fold and cut a square—only more involved due to the 36° prerequisite. The instructions should be self-explanatory.

As with the square, after students have obtained △GFH, they should not unfold the triangle. Point out that GF and GH are the folded sides of their triangle, as indicated by the broken lines. They represent two adjacent lines of reflection of the regular pentagon. The angle between them is the required 36°.

Show students the transparency of Master 11B. If you wish, have them make a five-pointed star or the ring of five paper dolls by starting with the folded pentagon (see △GFH), and then cutting away the shaded regions as per the pattern. The star involves only one cut, extending from a vertex between a folded and cut side to the midpoint of the folded side directly opposite. Otherwise, encourage them to make their own creation.

For a classic "golden star," known for its aesthetic perfect proportions, students must copy the pattern exactly. (Duplicate Master 11B as required.) The pattern has been drawn to scale, which you can demonstrate on the overhead by superimposing a folded triangle. Note that the cut does not extend precisely to the midpoint, but to a geometrically predetermined location. Of course, the star can be decorated with additional cuts.

STUDENT CHALLENGE

Try to create your own five-sided cutout. Start with another folded regular pentagon and then cut in a design. What did you make?

Paper & Scissors
PENTAGON

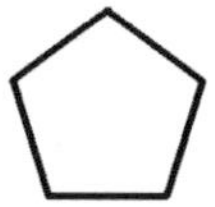

1 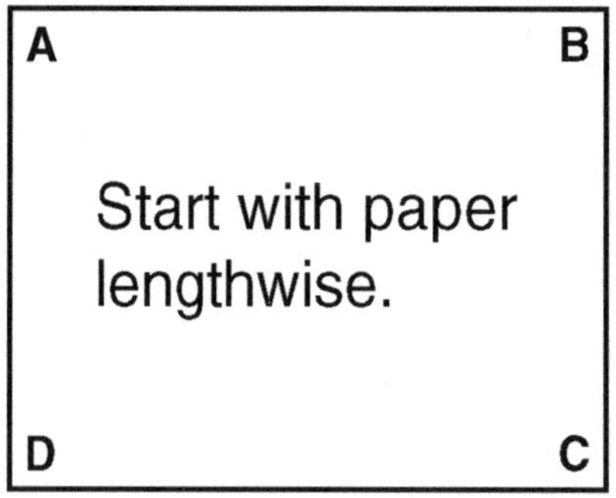 Start with paper lengthwise.

```
A          B
A          D
```
(corners labeled A, B at top; A, D at bottom) — *C*

2 Fold in half, left over right.

```
A          A
F ------- E
```
(D at top right, E at bottom right)

3 Match E to H, which is half the distance of F to A. Crease GJ.

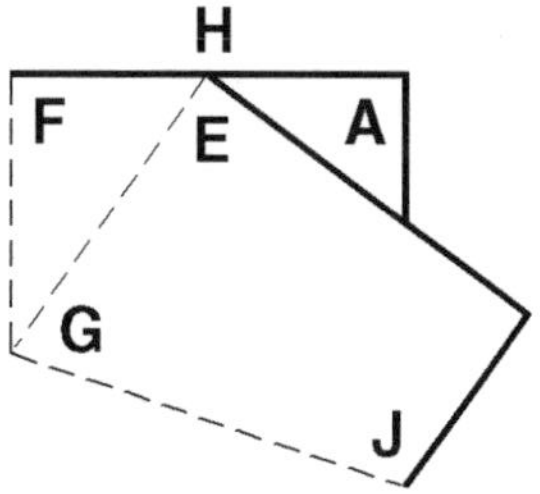

4 Bisect ∠EGJ by matching $\overline{GJ}$ to $\overline{GE}$. Crease on $\overline{GK}$.

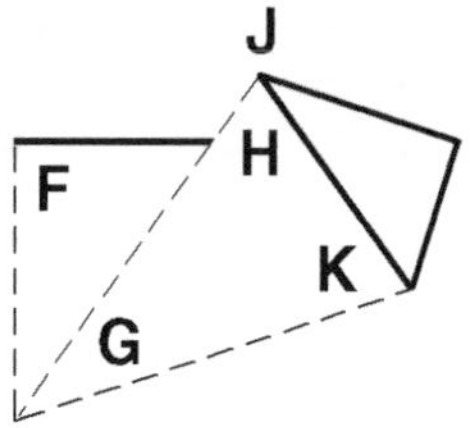

5 Turn over, flipping the bottom to the top.

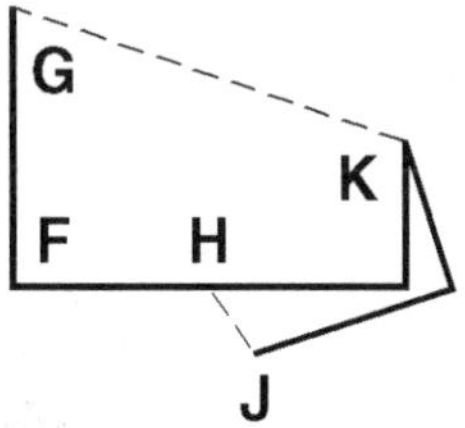

6 Bisect ∠FGK by matching $\overline{GF}$ to $\overline{GK}$. Crease on $\overline{GH}$.

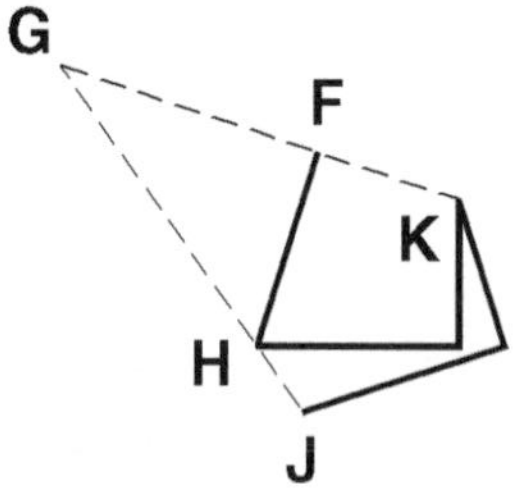

7 Cut on $\overline{HF}$.

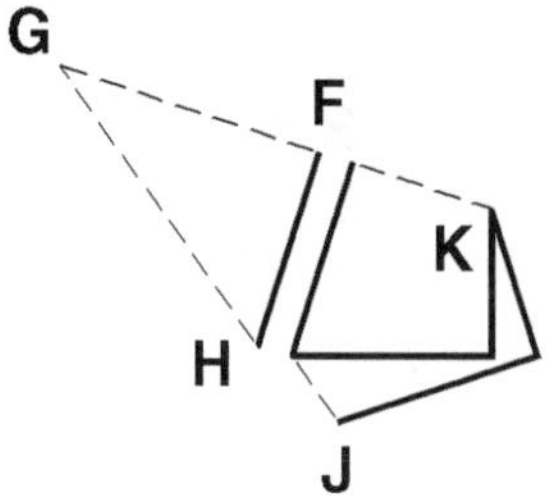

8 Save and unfold △GFH.

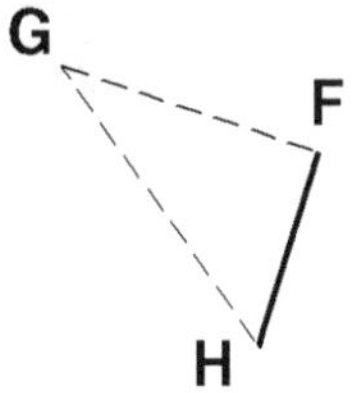

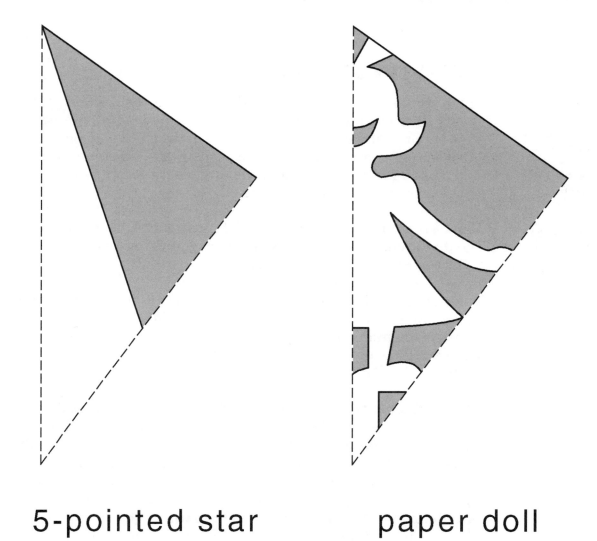

5-pointed star

paper doll

Paper and Scissors Hexagon

ACTIVITY SEQUENCE

MATERIALS

Master 9B, page 53 (handout
        from Activity 9)
Master 12A, page 65
        (transparency)
Master 12B, page 66
        (transparency)
Scissors, 1 per student
Blank sheets of lightweight
        paper, several per student
Overhead marker

Have students look back at the drawing of a regular hexagon with its six lines of reflection from Activity 9 (Master 9B). Mention that we proved earlier that the angle between two adjacent lines of reflection of a regular hexagon is $\frac{360°}{12} = 30°$. To create a paper and scissors regular hexagon, the angle between the folded sides of the eventual triangle must be 30°.

Show students the transparency of Master 12A. Have them follow the directions to create a regular hexagon using a piece of lightweight paper and scissors.

After students have obtained △GFJ, tell them to keep the triangle folded. Mention that GF and GJ are the folded sides of their triangle, as indicated by the broken lines. They represent two adjacent lines of reflection of the regular hexagon. The angle between them is the requisite 30°.

Show students the transparency of Master 12B. Snowflakes occur naturally as regular hexagons. The paper version at the bottom of the drawing was created by cutting the shaded regions into the folded hexagon at the top. The scalloped curve extending from vertex to vertex of the cut edge guarantees a regular hexagon appearance. Invite a student up to the overhead to add the lines of reflection to the snowflake with a marker. The connection between pattern and snowflake will soon become obvious.

STUDENT CHALLENGE

Create your own paper and scissors snowflake. Start with a folded regular hexagon and then cut in your own design. What did you make?

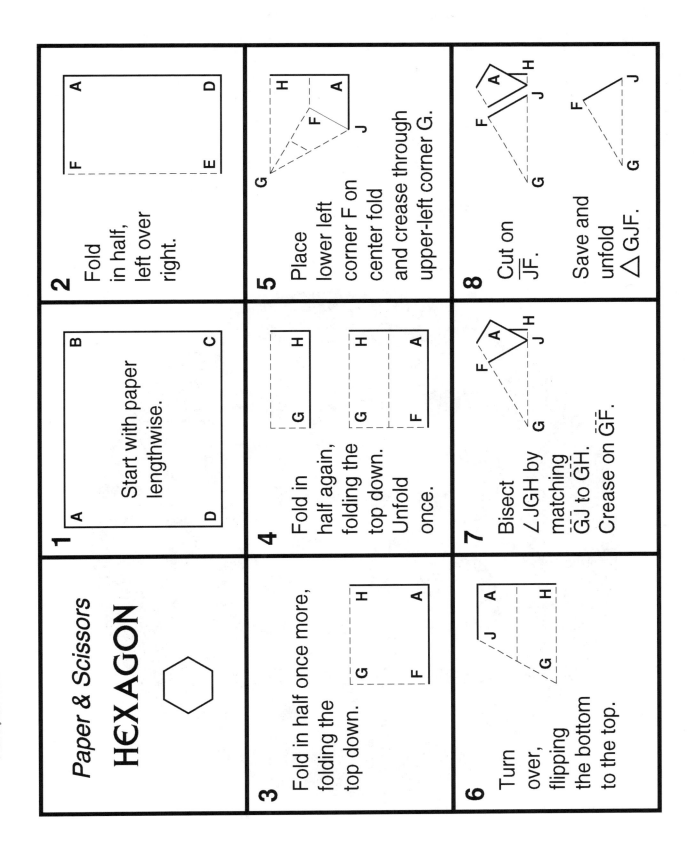

Paper & Scissors

HEXAGON

1 Start with paper lengthwise.

A · B
D · C

2 Fold in half, left over right.

A · D
F · E

3 Fold in half once more, folding the top down.

G · H
F · A

4 Fold in half again, folding the top down. Unfold once.

G · H

G · H
F · A

5 Place lower left corner F on center fold and crease through upper-left corner G.

H · A
F · J
G

6 Turn over, flipping the bottom to the top.

J · A
G · H

7 Bisect ∠JGH by matching $\overline{GJ}$ to $\overline{GH}$. Crease on $\overline{GF}$.

F · A
J · H
G

8 Cut on $\overline{JF}$.

Save and unfold △GJF.

F · A
J · H
G

F · J
G

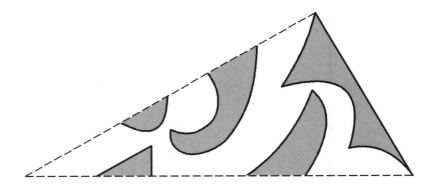

Angle Relations in Triangles and Quadrilaterals

ACTIVITY SEQUENCE

MATERIALS

Master 13A, page 69
    (transparency)
Sheet of paper
Rulers, 1 per student
Scissors, 1 per student
Blank sheets of paper,
    2 per student
Colored overhead marker

The angle measures in a triangle are interrelated. A simple demonstration reveals this relationship.

Show the transparency of Master 13A, masking out everything except the triangle with opaque paper or cardstock. Give each student a blank sheet of paper, a ruler, and scissors, then have them perform the following steps.

- Draw a very large triangle on the paper with the ruler and a pencil.

- Label the measures of the three angles of your triangle with the letters *a, b,* and *c* (see below).

- Cut out your triangle.

- Tear off each of the angles of your triangle. (Use large irregular tears to include the angle labels.)

- Align the three torn pieces so the three angles meet in a common point and are adjacent to one another. The three angles will form a straight angle, or 180°.

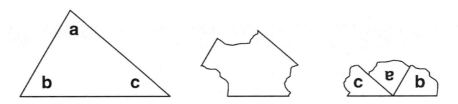

All of the angle *a*'s in the classroom are probably different, as are the *b*'s and *c*'s. However, the sum of the measures of the angles of each of the triangles is 180°. The sum of the measures of the interior angles of a triangle is always 180°. Reveal the first equation: $a + b + c = 180°$.

Have students follow these directions to find the sum of the angles of a quadrilateral.

- Draw a very large quadrilateral on paper with the ruler and a pencil.

- Label the measures of the four angles of your quadrilateral *a, b, c,* and *d.*

- Cut out your quadrilateral.

Reveal the quadrilateral on the transparency of Master 13A without revealing its equation. A quadrilateral has four angles. Add either diagonal with a colored marker. Tell students the angles can be subdivided into the angles of two triangles. Ask students the following question.

1. *If the sum of the measures of the interior angles of a triangle is always 180°, what is the sum of the measures of the interior angles of any quadrilateral?*

   The sum of the measures of the interior angles of any quadrilateral is 360°. Reveal the second equation on Master 13A: $a + b + c + d = 360°$.

To conclude, ask students to confirm this result as follows.

- Tear off each of the angles of your quadrilateral. (Use large tears as before.)

- Align the four torn pieces so the four angles meet in a common point and are adjacent to one another. The four angles will form a complete turn, or 360°.

STUDENT CHALLENGE

You have found the sum of the angle measures of a triangle and quadrilateral. Now try to find the sum of the angles of a pentagon and hexagon. How can you prove that your findings are correct?

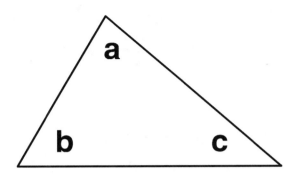

$$a \; + \; b \; + \; c \; = \; 180^{\circ}$$

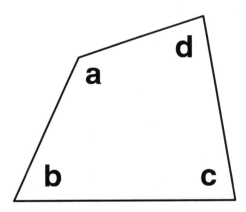

$$a \; + \; b \; + \; c \; + \; d \; = \; 360^{\circ}$$

Angle Measures in Regular Polygons

ACTIVITY SEQUENCE

Show the drawing of regular polygons from Activity 8 (Master 8B). Then discuss the following questions.

MATERIALS

Master 8B, page 46
    (transparency from
    Activity 8)
Master 14A, page 72 (handout)
Master 14B, page 73
    (transparency, optional)
Master 14C, page 74
    (transparency, optional)
Master 14D, page 75
    (transparency)
Calculators, 1 per student
Sheet of paper

1. *An equilateral triangle is a regular polygon. It has three equal angles and three equal sides. The sum of the measures of the interior angles of any triangle is 180° (from Activity 13). How large is the measure of each of the interior angles of an equilateral triangle?*

   $\frac{180°}{3} = 60°$

2. *A square is a regular polygon. It has four equal angles and four equal sides. The sum of the measures of the interior angles of any quadrilateral is 360° (from Activity 13). How large is the measure of each of the interior angles of a square?*

   $\frac{360°}{4} = 90°$

3. *Is there a simple way to determine the measure of each of the interior angles of any regular polygon?*

   Let students suggest answers, then give each student a copy of Master 14A—a table of interior angle measures.

Have students answer the following question, then fill in the first two angles on the table.

4. *What are the measures of each interior angle of a regular triangle and a regular quadrilateral (the first two results in the table)?*

   A regular triangle is an equilateral triangle. A regular quadrilateral is a square. The angles are 60° and 90°. Show the first two results on the transparency of Master 14D, but mask the other answers with opaque paper or cardstock.

Subject to student ability and time constraints, you can either derive the formula for the measure of the interior angle of a regular polygon using the transparencies of Master 14B and Master 14C, or present it as an established fact (angle measure $= 180° - \frac{360°}{n}$).

The approach to the derivation is visual; it emphasizes pattern and utilizes several results deduced earlier. Furthermore, it will produce a version of the formula that will be easy for the students to use.

Begin the derivation with a regular pentagon (show the transparency of Master 14B). If you wish, mask all but the drawing and its title with opaque paper or cardstock, then reveal the steps in stages. All five isosceles triangles are identical (congruent). Use a circle shape for the measure of each base angle and a diamond shape for the measure of each mirror angle (defined in Activity 8). The measure of each interior angle of the regular pentagon is "circle plus circle." When you reach the result in the box, ask students, the following question.

5. *What is the angle measure for a regular pentagon? Fill in the result on Master 14A.*

   $180° - 72° = 108°$ (Reveal the answer on the transparency of Master 14D.)

Repeat these steps for any regular polygon (show the transparency of Master 14C). Write the formula on the board so students may use it to complete the table.

STUDENT CHALLENGE

Use the formula and your calculator to complete the rest of the table on Master 14A. Note that a regular dodecahedron has 12 sides, that is, 2 more than 10. In two cases (regular heptagon and regular 42-gon), the result will not be a whole number. Express these results as mixed fractions.

[Students may use calculators equipped with a fraction button, or they may calculate on paper.] When appropriate, reveal the completed table for verification (Master 14D).

Polygon	Number of Sides and Angles	Measure of Each Interior Angle (Regular Polygons)
triangle	3	
quadrilateral	4	
pentagon	5	
hexagon	6	
heptagon	7	
octagon	8	
nonagon	9	
decagon	10	
dodecagon	12	
15-gon	15	
18-gon	18	
20-gon	20	
24-gon	24	
42-gon	42	

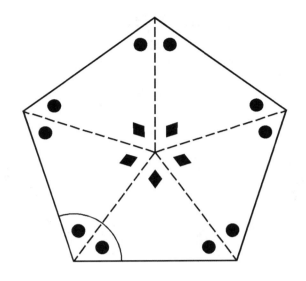

regular pentagon
5 sides

angle measure = ● + ●

● + ● + ◆ = 180°

● + ● = 180° − ◆

◆ = $\dfrac{360°}{5}$

angle measure = $180° - \dfrac{360°}{5}$

center

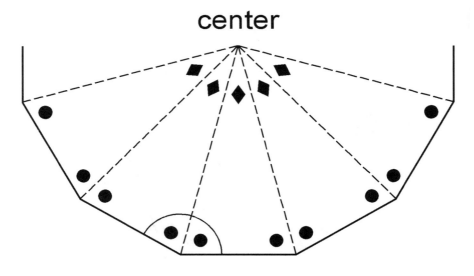

regular polygon
n sides

angle measure = ● + ●

● + ● + ◆ = 180°

● + ● = 180° − ◆

◆ = $\dfrac{360°}{n}$

angle measure = 180° − $\dfrac{360°}{n}$

Polygon	Number of Sides and Angles	Measure of Each Interior Angle (Regular Polygons)
triangle	3	$60°$
quadrilateral	4	$90°$
pentagon	5	$108°$
hexagon	6	$120°$
heptagon	7	$128\frac{4}{7}°$
octagon	8	$135°$
nonagon	9	$140°$
decagon	10	$144°$
dodecagon	12	$150°$
15-gon	15	$156°$
18-gon	18	$160°$
20-gon	20	$162°$
24-gon	24	$165°$
42-gon	42	$171\frac{3}{7}°$

ACTIVITY 15

The Regular Tessellations

MATERIALS

Master 8B, page 46
(transparency from
Activity 8)
Master 15A, page 79
(transparency and
handout)
Master 15B, page 80
(transparency)
Pieces of overlay, page 178
(transparency)
Overhead marker
Pattern blocks (optional)

VOCABULARY

tessellation
regular tessellation

ACTIVITY SEQUENCE

Prepare a transparency of the *Pieces of overlay* on page 178. Cut out the regular pentagons and regular octagons. Reserve the remnants for Activities 19 and 25.

Show the transparency of regular polygons from Activity 8 (Master 8B). Briefly discuss the following question.

1. *Suppose you wish to tile a floor and can use only one kind and one size of regular polygon tile. The tiles must meet side to side and vertex to vertex without any gaps or overlaps. Which regular polygons would be appropriate?*

   If time permits, provide students with pattern blocks of regular polygons and allow them to experiment. When done, show the transparency of Master 15A, the drawings of the three regular tessellations.

Even lacking hands-on experience with regular polygons, few students will fail to name the square. A few will suggest the equilateral triangle and the hexagon. Tell students they can tile the floor with congruent (identical) equilateral triangles, squares, or regular hexagons. The patterns are known as the *regular tessellations*. Each of the regular polygons therein is said to tessellate the plane. The word *tessellation* comes from the Latin "tessella," which was the small square stone tile used in ancient Roman mosaics.

Most likely, someone will name the regular pentagon. Invite a student up to the overhead projector, then produce the three regular pentagon cutouts. Challenge the student to position the pieces on the overhead so they have a common vertex and share pairs of sides. The class will soon realize that a gap is inevitable.

The measure of the interior angle of a regular pentagon is simply inappropriate. Regular pentagons will not tessellate the plane.

Repeat for the regular octagon. This time the students will discover that four octagons meeting side to side and vertex to vertex will surround a square hole (see Master 19A, page 105). Regular octagons alone will not tessellate the plane.

Show the transparency of the regular polygons equipped with the measures of their interior angles (Master 15B). The shaded polygons are those that tessellate. A regular polygon will tessellate the plane if and only if the measure of its interior angle in degrees divides 360° (a complete revolution) exactly. This is true only for the equilateral triangle, square, and regular hexagon. There are only three regular tessellations.

Mathematicians often use number names to refer to the regular tessellations. Write in the appropriate number names on Master 15A (shown here).

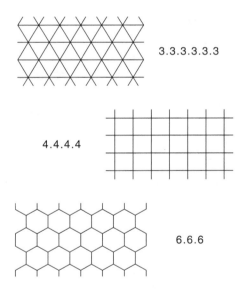

3.3.3.3.3.3

4.4.4.4

6.6.6

Ask students the following questions to identify and understand the use of the number names.

2. *Why do you think the symbol 4.4.4.4 is used for the tessellation of squares?*

Four four-sided regular polygons meet at (surround) every vertex.

3. *Explain the significance of the names 3.3.3.3.3.3 and 6.6.6.*

Six three-sided regular polygons or three six-sided regular polygons meet at (surround) every vertex in the corresponding regular tessellation.

LEARNING THE LANGUAGE

A *tessellation* is an arrangement of one or more shapes fitted together to cover a planar surface without any gaps or overlaps.

A *regular tessellation* is a tessellation of congruent (identical) regular polygons meeting side to side and vertex to vertex.

Provide each student with a copy of Master 15A to complete the following challenge.

STUDENT CHALLENGE

Mark the center (of rotation) of each complete polygon in the tessellation of equilateral triangles with a small dot. Connect these dots across the common sides to obtain a tessellation of regular hexagons. The tessellation of regular hexagons is said to be the *dual* of the tessellation of equilateral triangles.

Repeat for the tessellation of regular hexagons, marking the centers of each complete polygon with a dot, and connecting the dots across the common sides. You will obtain a tessellation of equilateral triangles. The tessellation of equilateral triangles and the tessellation of regular hexagons are duals of one another.

Repeat once again for the tessellation of squares. The tessellation of squares has a dual. What is it?

The answer is a tessellation of squares. The tessellation of squares is its own dual.

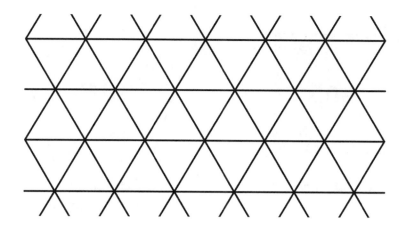

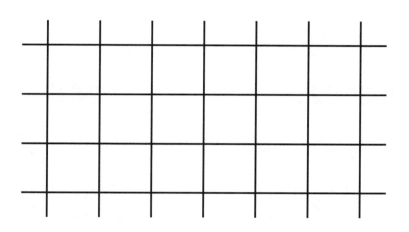

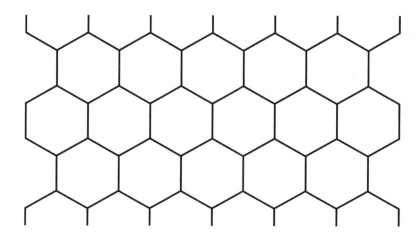

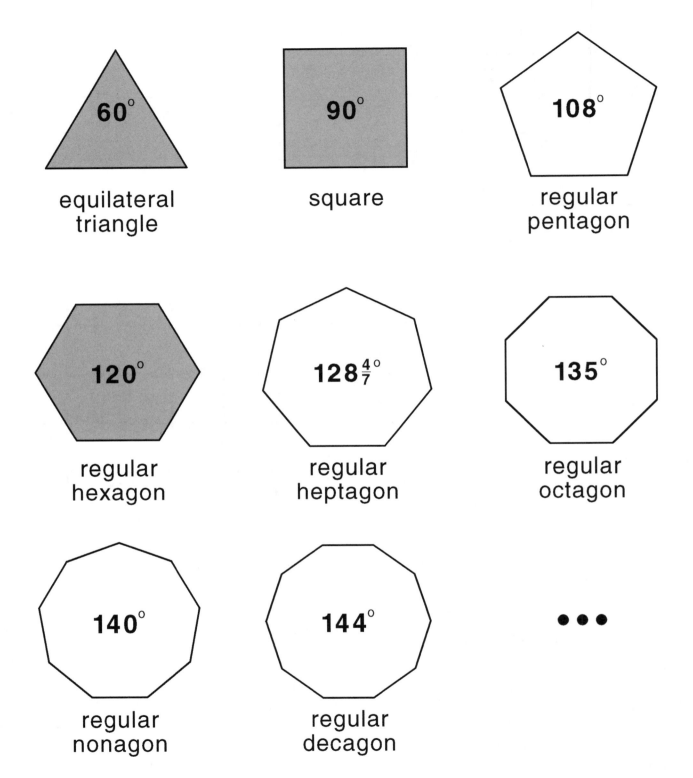

60°

equilateral
triangle

90°

square

108°

regular
pentagon

120°

regular
hexagon

128$\frac{4}{7}$°

regular
heptagon

135°

regular
octagon

140°

regular
nonagon

144°

regular
decagon

• • •

ACTIVITY 16

From Crackers to Soap Bubbles

MATERIALS

Master 16A, page 83
    (transparency)
Master 16B, page 84
    (transparency)
Master 16C, page 85
    (transparency)
6 round crackers
Bubble experiment materials
    (see page 82)

ACTIVITY SEQUENCE

Cracker manufacturers need to be aware of shapes that tessellate and those that do not. On the overhead projector, assemble a rectangular configuration of at least six touching circular crackers. If circular crackers are cut from a sheet of dough, there will be wasted material between the shapes as indicated by the gaps. To avoid waste, tessellating crackers are ideal. Show students the transparency of the box of regular hexagon crackers (Master 16A).

Briefly discuss the following question.

1. *What shapes do you think are ideal for cracker manufacturers to save dough?*

    Those that tessellate, including: equilateral triangles, squares, and regular hexagons. Other possibilities include rectangles and diamonds.

Honeycomb has a hexagonal framework. Show students the transparency of Master 16B. Bees have chosen the most efficient and economically-shaped packing container to store their honey. No other cell shape would be as economical of space and material. The hexagonal pattern is also the strongest from the standpoint of engineering design and is used in the construction of lightweight yet exceptionally strong doors. There are even hexagonal tessellations in your own eye!

Soap bubbles behave like honeycomb. When you blow a single soap bubble, the bubble's skin tries to hold the air inside it in a shape that contains the most volume for the least amount of material. That shape is a sphere. However, when soap bubble meets soap bubble, there are interesting consequences.

Pairs of students need the following materials to do an experiment with bubbles.

Materials: 9-inch aluminum pie plate

2 cups soapy solution (to each gallon of water, add $\frac{2}{3}$ cup of dishwashing liquid and 1 tablespoon of glycerine, available at your local pharmacy)

1 plastic spoon

1 jumbo straw

1 sheet $8\frac{1}{2}$-inch-by-11-inch clear acetate

This bubble solution works best if it is aged a day before use.

Have students pour the soapy solution into the pie tin. They should remove any bubbles on the surface of the liquid with the plastic spoon. Have students use the straws to blow a large bubble on the surface of the soapy solution. Add a second bubble.

2. *What happens to the common wall where the bubbles merge?*

   The wall is flat.

3. *Add a third bubble. What do you notice?*

   When three bubbles merge, their common walls meet at an angle of 120°.

4. *Fill the surface of the soapy solution with a single layer of large bubbles. Lay the sheet of acetate on top. What do you observe?*

   The cells form hexagons. The whole configuration is a hexagonal tessellation, albeit not a regular one, incorporating the classic 120° angles. Show the overhead of Master 16C. The individual cells look much like the cells of a beehive. As noted previously, bees try to be as efficient as possible when making honeycomb. They want to use the minimum possible amount of wax. A hexagonal tessellation is the ticket!

STUDENT CHALLENGE

Check your cupboards at home to see if you have any tessellating foods. Do your crackers, cookies, or nacho chips tessellate? Describe what you notice and explain why certain products tessellate and others do not.

ACTIVITY 16 MASTER 16B Investigating Patterns/**Symmetry and Tessellations**

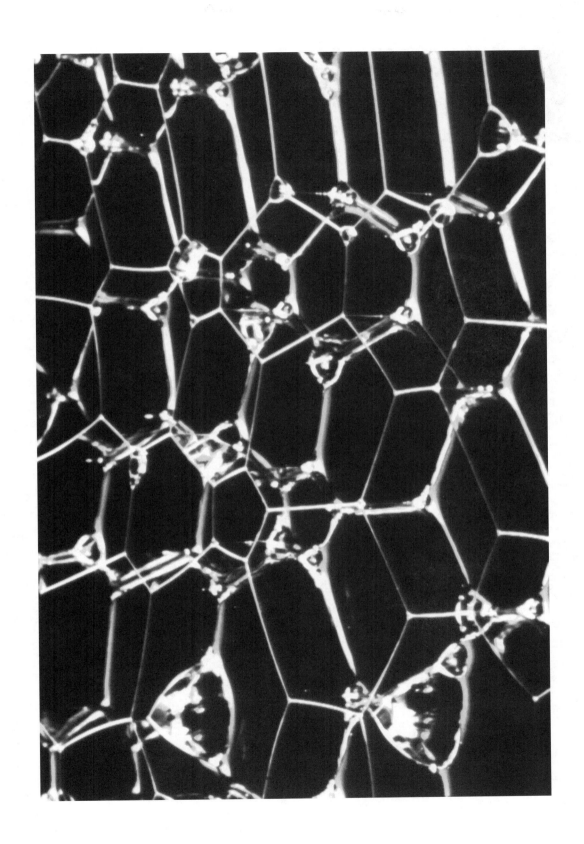

ACTIVITY 17

A Tessellation Kaleidoscope

Master 15A, page 79
     (transparency from
     Activity 15, 2 copies)
Master 17A, page 90
     (transparency)
Master 17B, page 91 (handout)
Master 17C, page 92 (handout)
Master 17D, page 93
     (transparency)
Hinged-mirror kaleidoscopes
     from Activity 8, 1 per pair
Mirrors, 1 per pair
Rubber bands, 2 per pair
Resealable plastic bags (to store
     mirrors and rubber
     bands), 1 per pair
Colored markers
Blank sheets of paper, 1 per
     student (optional)
Kaleidoscope materials (see
     the Student Challenge)

ACTIVITY SEQUENCE

Show students the transparency of the regular tessellations from Activity 15 (Master 15A). Make a duplicate of the transparency to use as an overlay. The regular tessellations are very symmetrical. They have reflectional symmetry (flip, then slide the overlay to align the polygons), rotational symmetry (rotate the overlay as required), and both translational and glide-reflectional symmetry (demonstrate with the overlay).

Provide each pair of students with a hinged-mirror kaleidoscope from Activity 8, a third mirror, and two elastic bands (one spare) in a resealable plastic bag. To begin, students should lay the hinged-mirrors flat, reflective surfaces down, and push the taped edges together to redefine the hinge. Next they stand the mirrors in the usual way with an angle of about 60° between them, and add the third mirror so its reflective surface faces that of the other original two. The three mirrors will form an equilateral triangle prism. Finally a student from each pair loops an elastic band around his/her thumbs and index fingers to form a square opening and stretches the elastic until the opening is large enough to envelop their prism (demonstrate). He/she lowers the band around the prism until about halfway down the mirrors and carefully releases the elastic (demonstrate). The mirrors will snap into a 60°- 60°- 60° prism kaleidoscope (see below).

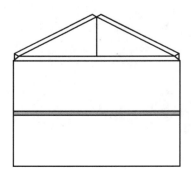

Place the transparency of Master 17A on the overhead. Then discuss the following question.

1. *Look at the transparency of Master 17A. Three of its numerous lines of reflection have been added to the tessellation of regular hexagons. What do you notice about these lines?*

    They intersect in the vertices of an equilateral triangle—similar to the triangle formed by your prism kaleidoscope. These vertices mark the centers (of rotation) of three regular hexagons. [If appropriate, you should recall the Student Challenge on duality from Activity 15.]

Provide each student with a copy of Masters 17B and 17C to answer these questions.

2. *Look at Master 17B. Do you recognize the figure at the top of the page? Lay the prism kaleidoscope on the figure so each mirror falls exactly on a side of the triangle. What happens?*

    The kaleidoscope forms a tessellation of regular hexagons.

3. *Look at the equilateral triangle at the bottom of Master 17B. It has been subdivided into three identical clown halves by three identical bold curves. Each curve starts at the triangle's center (of rotation) and proceeds to one of its vertices. This decorated triangle has rotational symmetry of order 3. Lay the prism kaleidoscope on the figure as before. What do you see?*

    The kaleidoscope generates a tessellation of clowns. This clown, seen in Activity 2, was designed so that it tessellates. A T-shirt imprinted with the tessellation appears in Activity 30. If you wish, have students color the clowns to dramatize the effect.

4. *What kind of symmetry do you see in the clown tessellation?*

    The tessellation has reflectional, rotational (order 3), translational, and glide-reflectional symmetry.

Now have students look at Master 17C. Tell them these decorated equilateral triangles are reconstructions of two figures by M. C. Escher. (Activity 24 details Escher's tessellations.) Have students study the three fish halves in the top figure. Like the clown halves, these were formed by drawing three identical curves from the triangle's center to each of its vertices. Again, the decorated triangle has rotational symmetry of order 3.

If the students study the decorated triangle at the bottom, they will find that three distinct rather than three identical curves were drawn from the center of the triangle to each of its vertices to create the bat, reptile, and fish halves.

Have students color each creature on the handout a different color. Escher colored the fish red, the reptiles blue, and the bats yellow in his version. Have students lay their prism kaleidoscopes on the decorated triangles to create the tessellating art.

Show students the transparency of Master 17D—a greyscale version of Escher's bat/reptile/fish tessellation. A colored version appears in *Visions of Symmetry* (see Bibliography).

As an optional exercise, have students trace any of the four equilateral triangles on blank paper, then decorate the triangle before looking at it with the prism kaleidoscope. For an Escher-like effect, students should begin with either identical or distinct curves emanating from the triangle's center and proceeding to each vertex.

STUDENT CHALLENGE

Almost everyone has looked through a kaleidoscope. The word *kaleidoscope* comes from a Greek phrase meaning "to view a beautiful form." Sir David Brewster invented the toy in 1816. But how many people know how it is made or how it works its magic?

Make your own kaleidoscope using the materials listed here.

Materials: clean potato chip can with translucent plastic lid
knitting needle or sharp scissors
10-inch-by-10-inch piece of colorful contact paper
8-inch-by-10-inch piece of silvered mylar
8-inch-by-10-inch piece of cardboard, about $\frac{1}{16}$-inch thick
stylus
ruler
masking tape
3-inch-by-3-inch piece of heavy clear acetate (such as the lid to a box of greeting cards)
$\frac{3}{8}$-inch-by-10-inch cardboard spacer strip, about $\frac{1}{16}$-inch thick
paper clips, rhinestones, translucent plastic beads, or any other small colorful translucent items

1. Turn the can upside down. Use a knitting needle or pair of sharp scissors to puncture an eyehole about $\frac{1}{4}$-inch in diameter

in the center of the metal end. (If the eyehole is irregular, use a one-hole paper punch to cut a hole in the center of a large circular label, then stick the shape over the opening.)

2. Decorate the outside of the can with the contact paper.

3. Glue the silvered mylar to the piece of cardboard. Use a stylus and ruler to cut the covered cardboard into three "mirrors," each 85% of the interior diameter of the can in width and $\frac{3}{8}$-inch less than the interior depth of the can in length. (Plastic or glass mirrors, cut to the correct size, are more costly and marginally more effective alternatives.)

4. Lay the mirrors reflective side down, then line them up long sides together with a $\frac{1}{16}$-inch gap between them. Hinge the long sides of the mirrors together with two long strips of masking tape, trimming the ends of the tape.

5. Fold the mirrors into an equilateral triangle prism with the reflective side on the inside. Use additional strips of masking tape to hold the configuration in place. (One edge of each mirror should just meet the edge of the next mirror around the prism.)

6. Fit the mirrors into the can.

7. Cut a circle from the acetate precisely the size of the can opening, then lay the circular shape on top of the triangular prism. An optional thin ribbon of glue on the top edge of the mirrors will fix the acetate in place.

8. Line the inside of the can between the acetate circle and the top rim of the can with the cardboard spacer strip, trimming as required, and joining the ends with a small piece of masking tape.

9. Fill the upper compartment about half full with the paper clips, rhinestones, and other translucent objects collected, then put on the plastic lid. Do not use too many objects or they will not tumble well and may block out too much light.

10. Hold the kaleidoscope up to the light, look into the eyehole, and slowly rotate the can. The three mirrors will symmetrically reflect the objects under the lid and produce the kaleidoscope effect.

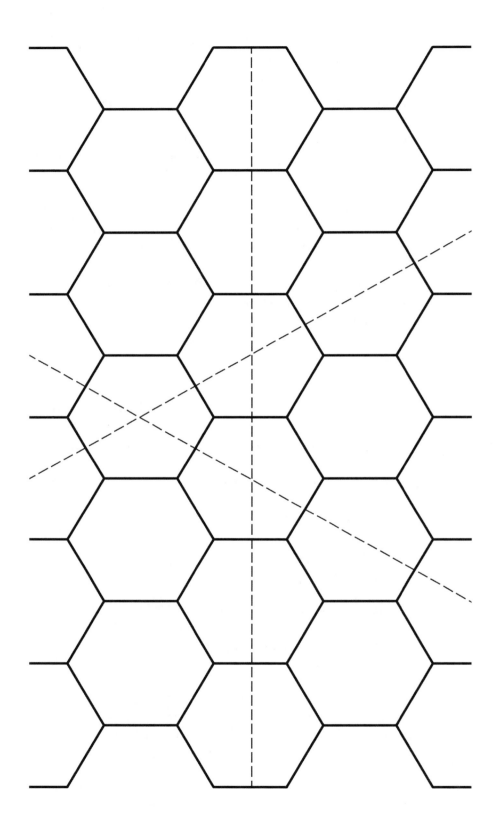

ACTIVITY 17 MASTER 17A Investigating Patterns/**Symmetry and Tessellations**

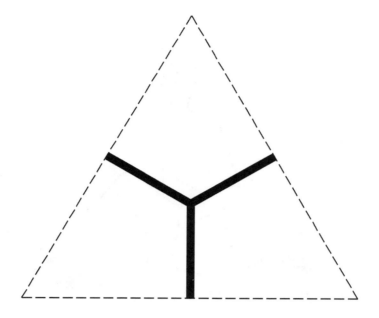

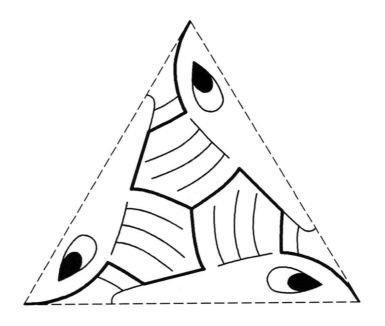

ACTIVITY 17 MASTER 17C Investigating Patterns/**Symmetry and Tessellations**

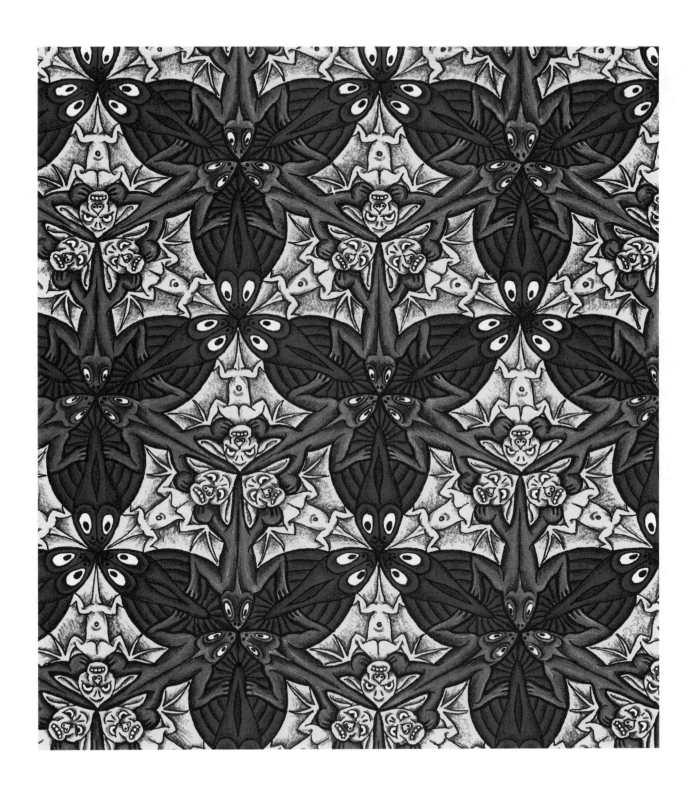

Tessellating with Irregular Polygons

ACTIVITY SEQUENCE

MATERIALS

Master 18A, page 97
        (transparency, 2 copies)
Master 18B, page 98
        (transparency and copy
        on cardstock)
Master 18C, page 99
        (transparency, 2 copies)
Master 18D, page 100
        (transparency, 2 copies)
Master 18E, page 101
        (transparency, 2 copies)
Scissors, 1 per student

VOCABULARY

midpoint rotation
kite quadrilateral

Show students the transparency of Master 18A. Explain that any rectangle will tessellate the plane; and any parallelogram will as well. If the parallelogram tiles can be flipped, not the case with most ceramic ones, they can be used to make more than one distinctive pattern (point to the two tessellations at the bottom of the transparency).

Overlay a duplicate of the transparency to investigate the symmetries of the tessellations. Flip, turn, and slide the overlay as required. The tessellation of parallelograms in the lower-right corner is particularly interesting. It has vertical glide-reflectional symmetry without vertical reflectional symmetry.

There are many tessellating irregular polygons. Have students suppose that in their search for the perfect tile, they came across a box of identical scalene quadrilateral tiles (show the transparency of Master 18B). The sides of this quadrilateral are of different lengths, and its angles are of no specific measure. It seems such a polygon will not tessellate the plane . . . or will it?

Duplicate Master 18B on cardstock. Have students cut out the 12 quadrilaterals, then challenge them to see if the polygon will tessellate. Emphasize that the quadrilaterals must meet side to side and vertex to vertex. Require that at least one cutout be totally surrounded by the others. (Point out that the measures of the various congruent (identical) angles have been indicated by letters of the alphabet).

When appropriate, reveal the solution to the puzzle (transparency of Master 18C), but mask the equation at the bottom of the page. Point out that the four angles of the quadrilateral surround each interior vertex of the tessellation. Ask students the following questions.

1. *Why is there no gap in the solution?*

   Because the sum of the measures of the interior angles of any quadrilateral is always 360°, a complete revolution (see Activity 13). Reveal the equation on the bottom of Master 18C.

Use a duplicate of the transparency as an overlay to show that the pattern has translational symmetry. Then align the drawings on the two layers, select the midpoint of any side of either of the two interior quadrilaterals in the tessellation, and rotate the overlay 180° about that midpoint. The coincidence of quadrilaterals will be maintained. If the pattern were extended indefinitely, it would coincide with itself exactly. The tessellation has rotational symmetry.

2. *The midpoint of any side of any quadrilateral is a center of rotation. What is the angle of rotation?*

   180°

3. *What is the order of rotation?*

   2

Using the transparency of Master 18A, show students how any parallelogram can be subdivided into two congruent triangles, usually scalene, by adding either of its diagonals. Since any parallelogram will tessellate the plane, so will any scalene triangle. Show the overhead of Master 18D. The midpoint of any side of any triangle is a center of rotation of the tessellation. The angle of rotation is 180°, and the order of rotation is 2, the same as the scalene quadrilateral tessellation.

Kites are curious tessellating quadrilaterals. Explain to students that a kite is any quadrilateral for which at least one diagonal is a line of reflection. In Master 18D, the kite tessellation's line of reflection is horizontal. Adding this line would divide any constituent quadrilateral into two congruent (identical) scalene triangles (demonstrate on transparency). All four angles of the quadrilateral surround every vertex.

Point out to students that the tessellation has vertical glide-reflectional symmetry, but lacks vertical reflectional symmetry.

Use a duplicate of the transparency as an overlay. Flip it left to right, then slide it until the tessellations of kites coincide.

Irregular hexagons can tessellate like their regular cousins. Show the transparency of Master 18E. If a hexagon has three sets of parallel and equal opposite sides, it will tessellate by translation. This is true of the first hexagonal tessellation illustrated (demonstrate using the duplicate transparency as an overlay). If you slide the overlay in any of three appropriate directions (perpendicular to the pairs of sides), the hexagons on the upper layer will slide into place with those on the lower layer.

A hexagon can tessellate by rotation as well (point to the tessellation at the bottom of the transparency). If every second (alternate) angle in the hexagon has a measure of 120° and the respective arms of those angles are equal in length, the corresponding tessellation will have rotational symmetry of order 3. The tessellation will have three distinct centers of rotation (namely the vertices of the 120° angles).

Demonstrate by using the duplicate as an overlay on the other hexagonal tessellation. Select any one of the three distinct centers of rotation as a pivot point, then invite students to the overhead to repeat the procedure with the other centers. [They should seek out interior vertices at which three congruent (identical) angles meet. The three line segments emanating from the common vertex will be equal in length.] In Activity 29, it will be revealed that such a tessellating hexagon forms the basis of Escher's reptile (originally encountered in Activity 7).

LEARNING THE LANGUAGE

In *midpoint rotation,* a shape rotates 180° around the midpoint of one of its sides.

A *kite quadrilateral* is a quadrilateral with at least one diagonal that is a line of reflection.

STUDENT CHALLENGE

There are other tessellating quadrilaterals besides squares, rectangles, parallelograms, and kites. Can you think of characteristics they must have to ensure that they meet side to side and vertex to vertex?

Students may use *TesselMania!®* (see Activity 30) to reveal all tessellating quadrilaterals. A demo version of the software will suffice for this purpose (see Appendix).

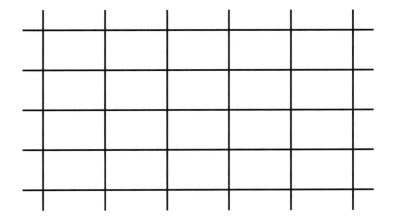

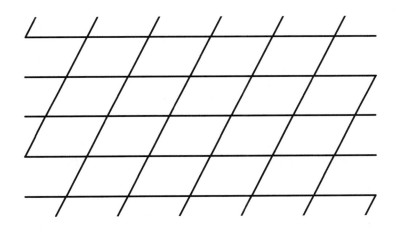

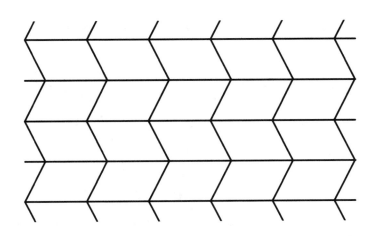

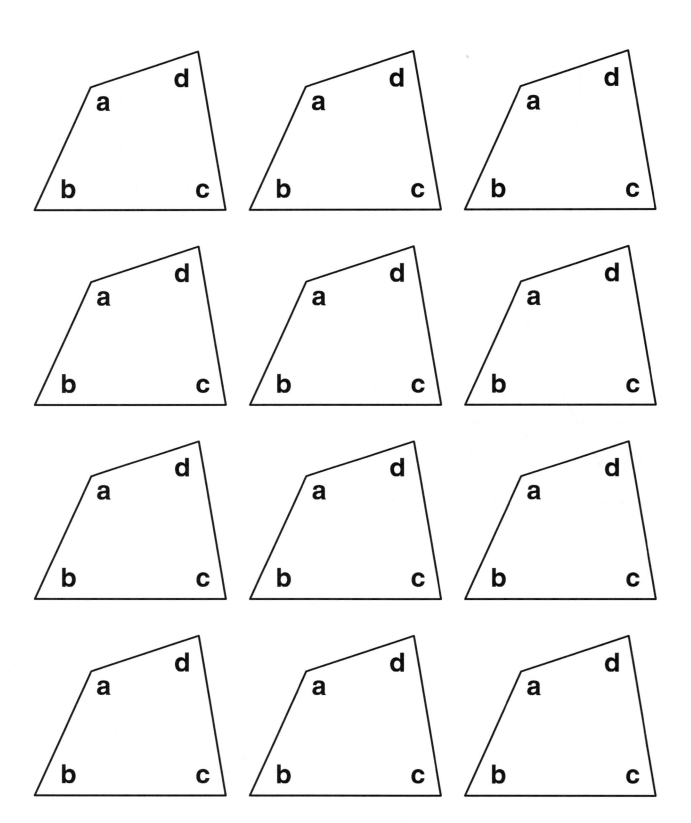

ACTIVITY 18 MASTER 18B Investigating Patterns/**Symmetry and Tessellations**

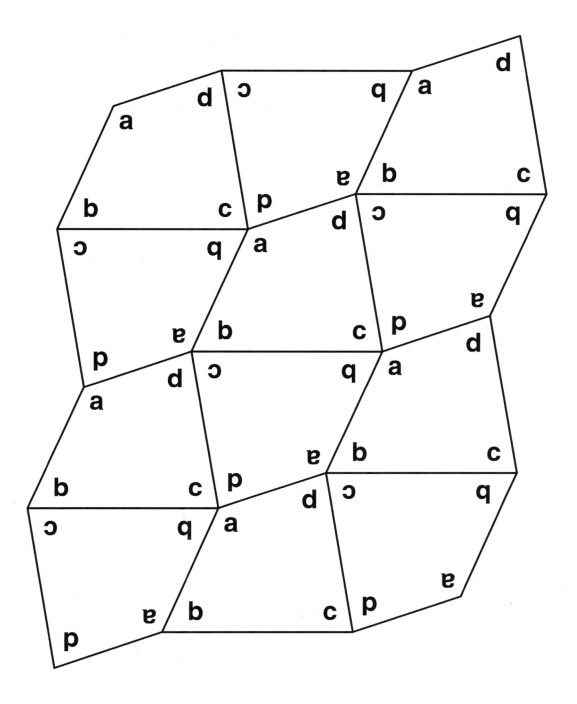

$$a + b + c + d = 360°$$

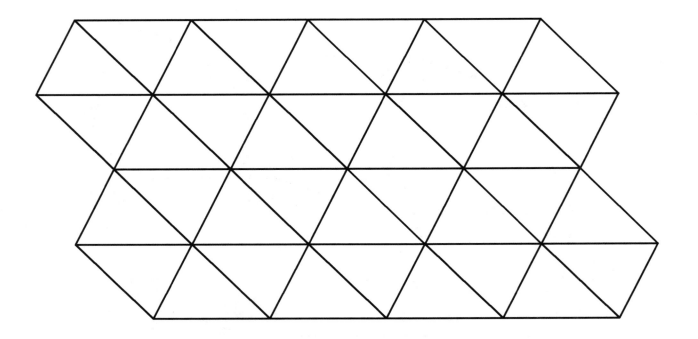

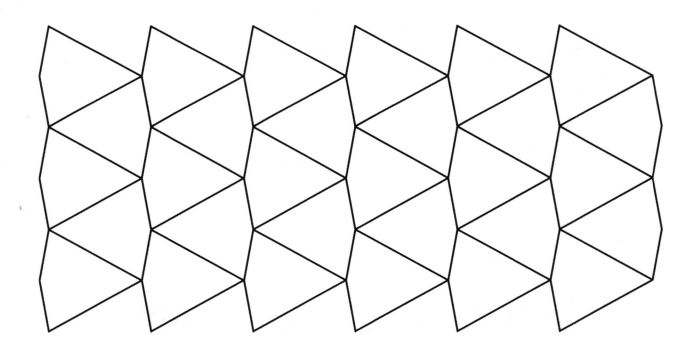

ACTIVITY 18 MASTER 18D Investigating Patterns/**Symmetry and Tessellations**

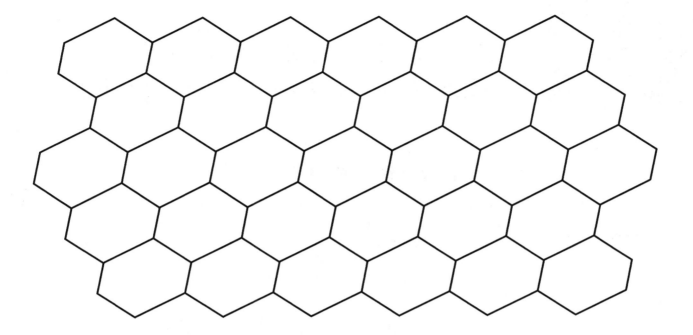

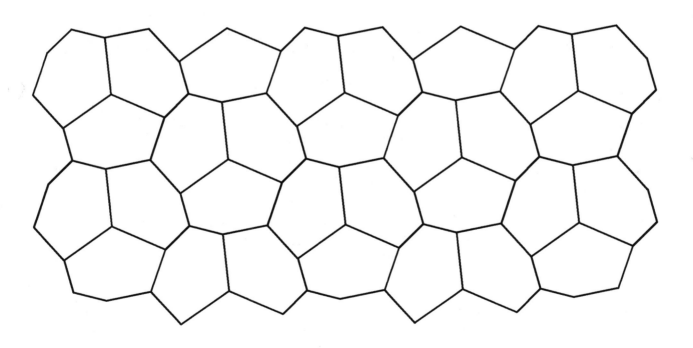

Surrounding a Point with Regular Polygons

ACTIVITY SEQUENCE

Use scissors to extract the equilateral triangle from the remnants of the transparency of *Pieces of overlay* (used in Activity 15). To do this, cut out a 1-inch square to completely surround the triangle. The excess acetate around the triangle is transparent, and the larger piece is easier to handle than if you cut out just the small triangle.

In Activity 15, students learned that regular octagons will not tessellate the plane. They also discovered that four regular octagons meeting side to side and vertex to vertex will surround a square hole. It appears that a combination of regular octagons and squares will tessellate the plane.

Show the transparency of Master 19A. Explain that this tessellation involves more than one kind of regular polygon, and so it is not a regular tessellation. However, it is repetitive in that every vertex has two regular octagons and a square surrounding it. Have students answer the following question.

1. *Can you suggest an appropriate number name for the regular octagon and square tessellation?*

    4.8.8—One four-sided polygon and two eight-sided polygons meet at (surround) every vertex.

The tessellation of regular octagons and squares is familiar. It often appears in modern linoleum and ceramic tile patterns (show the transparency of Master 19B). An actual piece of linoleum tile would be a visual asset.

Are there other combinations of more than one kind of regular polygon that will generate repetitive patterns? Many texts have students experiment with cutouts of regular polygons of a fixed side length. The exercise can lead to erroneous results.

Show the transparency of Master 19C. The large regular polygon has 20 sides (write 20 in the 20-gon), and the smaller one is a regular octagon (write 8 in the octagon). It would appear that if you add an equilateral triangle to the drawing, you can surround either of the shared vertices exactly (place the equilateral triangle overlay so the vertices of the three polygons touch). Ask the class if the pieces fit together exactly, or if there is a gap between the polygons? Does the combination 3.8.20 fill the space around a point exactly?

Remind students of the formula "angle measure = $180° - \frac{360°}{n}$." Display the transparency of Master 14D (Master 14A with answers) and have students answer the following questions using their own, filled-in copies of Master 14A.

2. *How large is the interior angle in a regular 20-gon?*

    162°

3. *How large is the interior angle in a regular octagon?*

    135°

4. *How large is the interior angle in an equilateral triangle?*

    60°

5. *What is the sum of the three angle measures?*

    162° + 135° + 60° = 357°

6. *The combination 3.8.20 will not fill all the space around a point exactly, that is, the combination will not surround a point. How can you tell that this is true?*

    It fails to surround a point because there is a gap of 360° − 357° = 3°.

7. *Using Master 14A and a calculator, determine what combinations of the measures of regular polygon angles will surround a point, that is, total 360°. As before, use the appropriate number name for the various polygons. There are only 17 such angle combinations possible, including 3.3.3.3.3.3, 4.4.4.4, 6.6.6, and 4.8.8 discovered previously, so you have only 13 more to find.*

    When appropriate, collect the results in writing or in a free-for-all, then show the solution (transparency of Master 19D). There are 21 combinations pictured. The additional four arrangements are made by placing some of the polygons in a different circular or cyclic order, as in 3.6.3.6 and 3.3.6.6.

STUDENT CHALLENGE

Visit a ceramic tile store and look for square sheets containing regular polygon tiles. [Most polygonal tiles are fixed to square sheets for ease of installation.] Look for combinations of regular polygons that surround a point exactly. The combination 4.4.4 is sure to predominate, but can you find 6.6.6 or 3.3.3.3.3.3? How about 4.8.8? Can you find other combinations encountered in this activity? Does the pattern continue throughout the sheet?

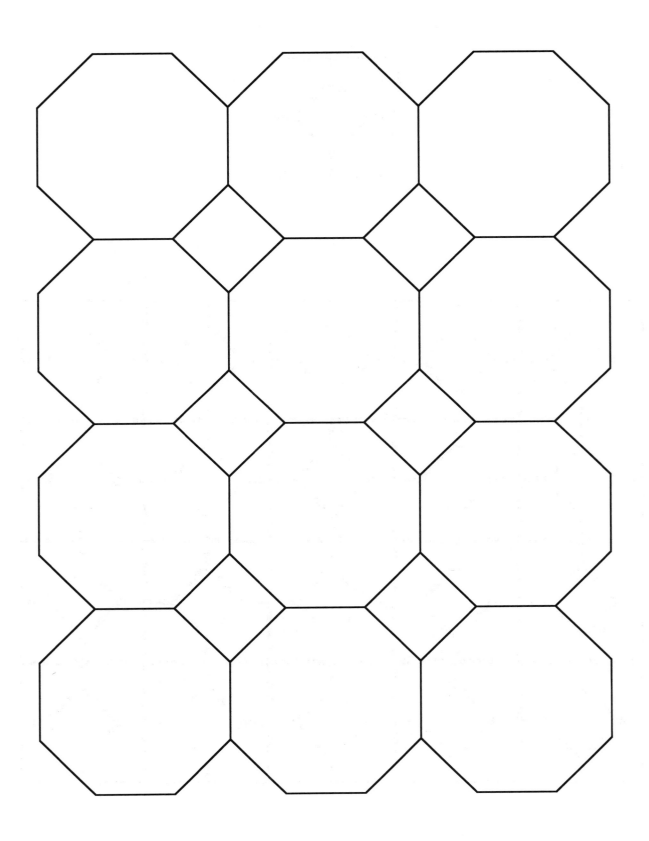

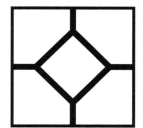

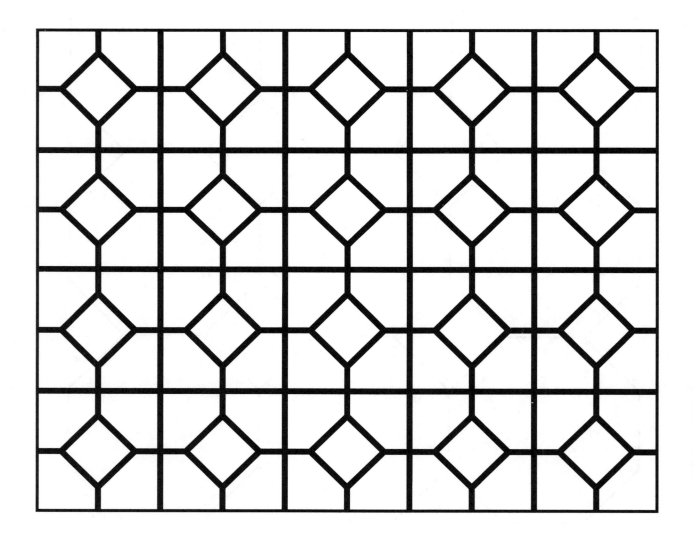

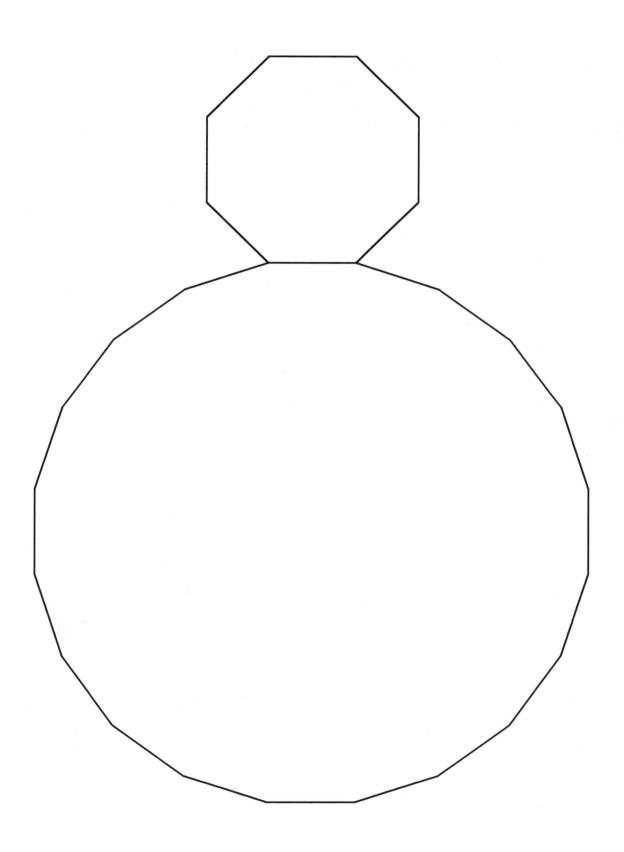

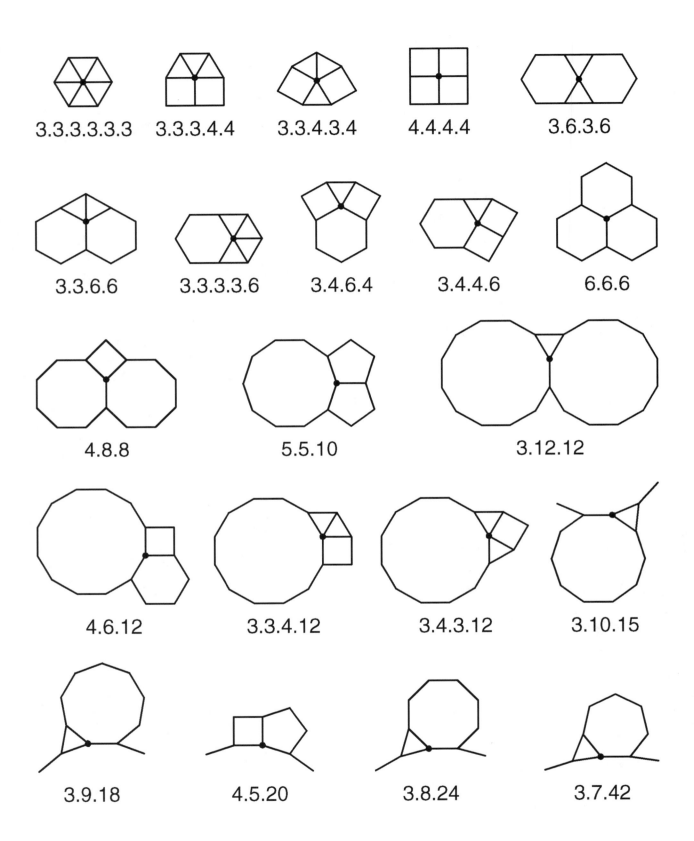

3.3.3.3.3.3   3.3.3.4.4   3.3.4.3.4   4.4.4.4   3.6.3.6

3.3.6.6   3.3.3.3.6   3.4.6.4   3.4.4.6   6.6.6

4.8.8   5.5.10   3.12.12

4.6.12   3.3.4.12   3.4.3.12   3.10.15

3.9.18   4.5.20   3.8.24   3.7.42

ACTIVITY
20

The Semiregular Tessellations

MATERIALS

Master 19D, page 108
   (transparency from
   Activity 19)
Master 20A, page 111 (handout)
Master 20B, page 112
   (transparency)
Pattern blocks (optional)

VOCABULARY

semiregular tessellation

ACTIVITY SEQUENCE

Now that students have identified the 21 arrangements of regular polygons that will surround a point (fill the space around a point), have them investigate which of these can be extended indefinitely.

Provide each student with a copy of Master 20A. Show the transparency of Master 19D, then ask the students the following questions.

1. *Three of the 21 combinations of regular polygons can be continued indefinitely to produce a regular tessellation. Which ones are they?*

   3.3.3.3.3.3, 4.4.4.4, and 6.6.6

Filling the space around a point guaranteed a tessellation when using congruent (identical) regular polygons, but it is no guarantee when using more than one of these shapes. If you have pattern blocks, the students can use them to discover that some of the remaining 18 combinations can be continued, and some cannot.

Only 8 of remaining 18 combinations can be continued indefinitely. These have been assembled on Master 20A. Each is a *semiregular tessellation*. Notice how in each tessellation an identical set of regular polygons surrounds every vertex.

2. *The classic linoleum tile pattern from the previous activity is a semiregular tessellation. Which pattern is it?*

   The one in the third row on the left.

3. *In Activity 19, we gave this pattern the number name 4.8.8. Explain the reasoning for this.*

   Precisely 1 square and 2 octagons surround every vertex in the tessellation.

4. *Can you provide number names for each of the other semiregular tessellations? Write in appropriate number names on Master 20A. Note that the usual numbering system uses a cyclic or circular order, that is, the polygons are listed in sequence as you proceed in a small circle around any vertex.*

Reveal the answers using the transparency of Master 20B.

5. *These are the usual number names. Notice that the numbering system lists the greatest number of smaller numbers first, although that is not essential here. Study the number names carefully. Two of the patterns involve the same regular polygons in a different cyclic order. Which patterns are they?*

3.3.3.4.4 and 3.3.4.3.4

LEARNING THE LANGUAGE

A *semiregular tessellation* is a tessellation of regular polygons of more than one kind meeting side to side and vertex to vertex in such a way that the same polygons, in the same cyclic (circular) order surround every vertex.

STUDENT CHALLENGE

The three regular and eight semiregular tessellations are sometimes called the Archimedean tessellations. Research Archimedean tessellations and explain the origin of the name.

Three different patterns consisting of equilateral triangles and squares are shown below. One of them is not a semiregular tessellation because each vertex is not surrounded by the same regular polygons. Which one is it?

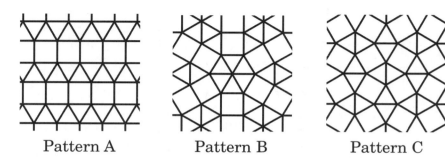

Pattern A   Pattern B   Pattern C

The answer is Pattern B.

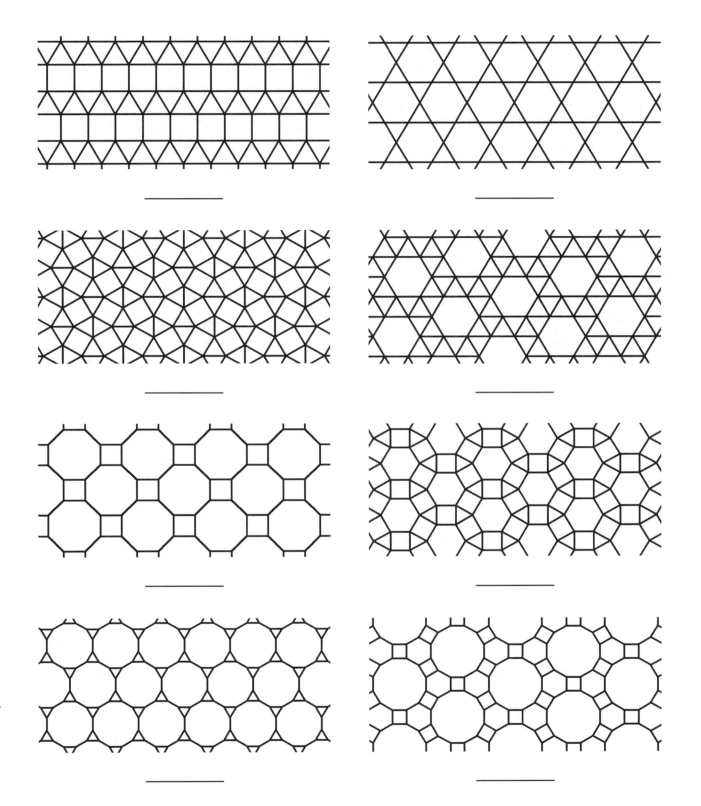

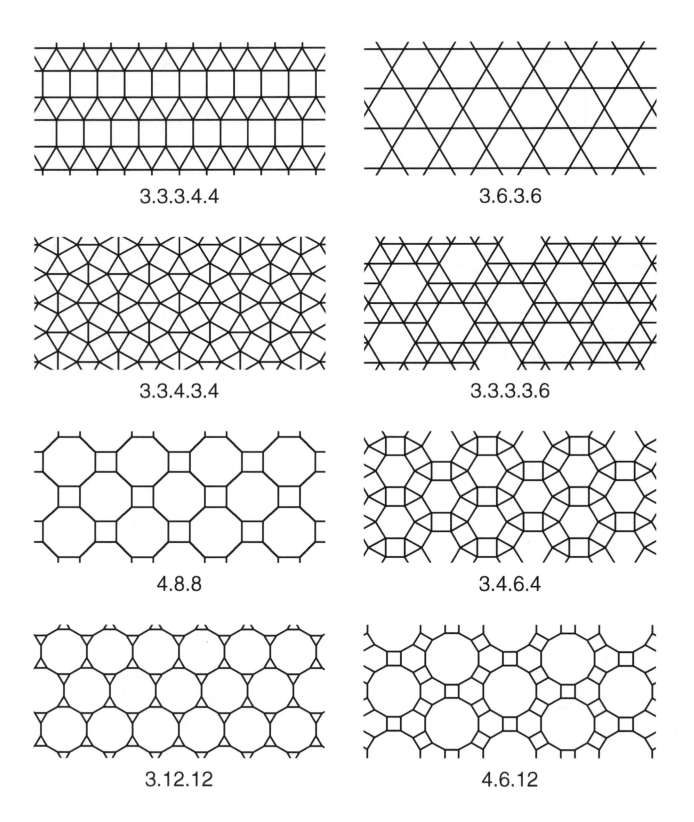

3.3.3.4.4

3.6.3.6

3.3.4.3.4

3.3.3.3.6

4.8.8

3.4.6.4

3.12.12

4.6.12

A Semiregular Tessellation Kaleidoscope

ACTIVITY SEQUENCE

MATERIALS

Master 20B, page 112
    (transparency from
    Activity 20)
Master 21A, page 114 (handout)
Master 21B, page 115 (handout)
Hinged-mirror kaleidoscopes
    from Activity 8, 1 per pair
Mirrors, 1 per pair
Rubber bands, 1 per pair
Resealable plastic bags (to store
    mirrors and rubber
    bands), 1 per pair
Colored markers

Show the semiregular tessellations with their usual number names from Activity 20 (Master 20B).

Have students reassemble their prism kaleidoscopes (one per pair of students) as in Activity 17. These reflective devices can be used to generate at least four of the semiregular tessellations. This is possible because, like the regular tessellations, the semiregular tessellations are very symmetrical. The only requirement: three of the various lines of reflection of the tessellation must outline an equilateral triangle (see Master 17A).

Give each student a copy of Master 21A and Master 21B. Have students answer the following questions.

1. *Lay your prism kaleidoscope on each of the triangle outlines on Master 21A. Which semiregular tessellations are generated? Give their number names.*

    3.6.3.6 and 3.12.12

2. *Now lay your prism kaleidoscope on each of the triangle outlines on Master 21B. Which semiregular tessellations are generated? Give their number names.*

    3.4.6.4 and 4.6.12

STUDENT CHALLENGE

Color one of the triangles from Master 21A or 21B. Use your prism kaleidoscope to generate your design. Isn't symmetry wonderful!

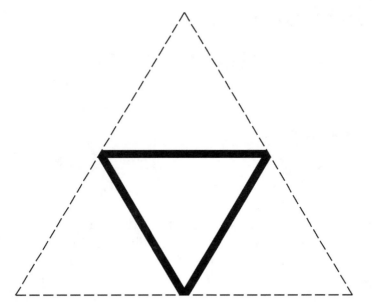

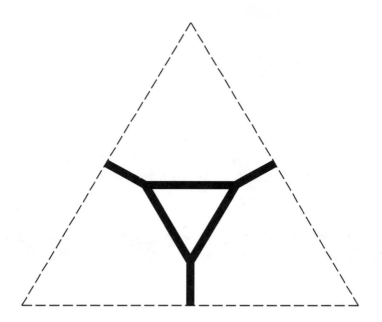

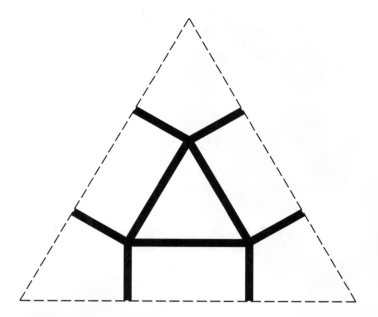

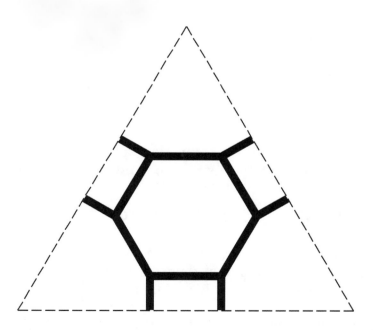

Investigating Patterns/**Symmetry and Tessellations**   ACTIVITY 21   MASTER 21B

Demiregular Tessellations

ACTIVITY SEQUENCE

In semiregular tessellations, all vertices are surrounded by identical combinations of regular polygons.

Hand out copies of Master 22A. Using the transparency, point out to students that the tessellation is made from regular polygons and that its pattern can extend indefinitely. The pattern is not semiregular because two distinct combinations of polygons surround its vertices. Ask students the following question.

1. *What are the number names of these distinct combinations?*

   3.4.6.4 and 3.3.4.3.4. The tessellation is said to be demiregular.

Hand out copies of Master 22B and have students answer the following question.

2. *Three distinct combinations of regular polygons surround the vertices of this tessellation. What are their number names?*

   3.3.3.3.3.3, 3.3.4.3.4, and 3.3.3.4.4

This completes our exploration of which regular polygons tessellate, both by themselves and in combinations.

LEARNING THE LANGUAGE

A *demiregular tessellation* is a tessellation of regular polygons meeting side to side and vertex to vertex in such a way that more than one distinct combination of polygons surround its vertices.

MATERIALS

Master 22A, page 117
 (transparency and
 handout)
Master 22B, page 118 (handout)
Colored markers

VOCABULARY

demiregular tessellation

STUDENT CHALLENGE

Color the two demiregular tessellations (Masters 22A and 22B) in a systematic way to emphasize their distinctive vertices.

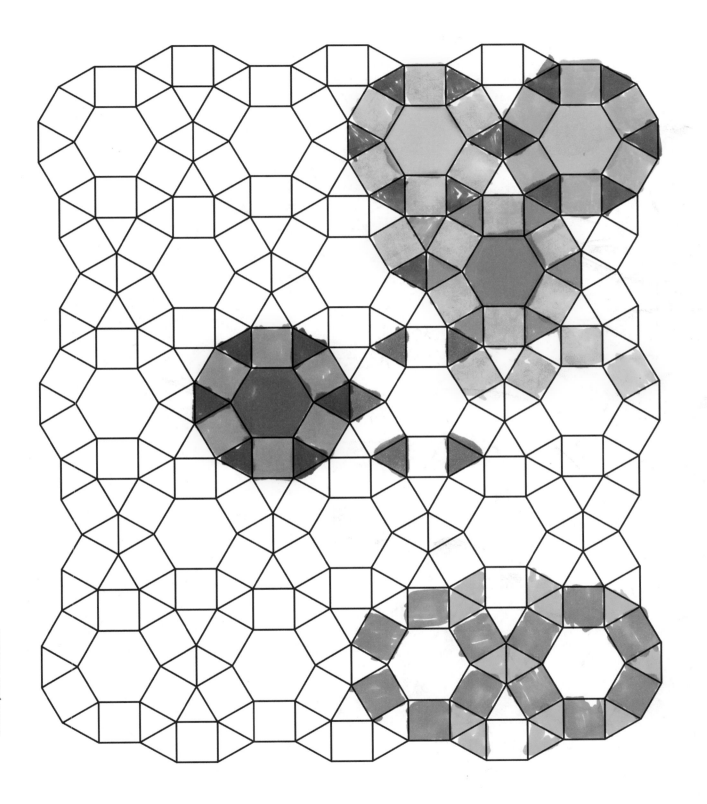

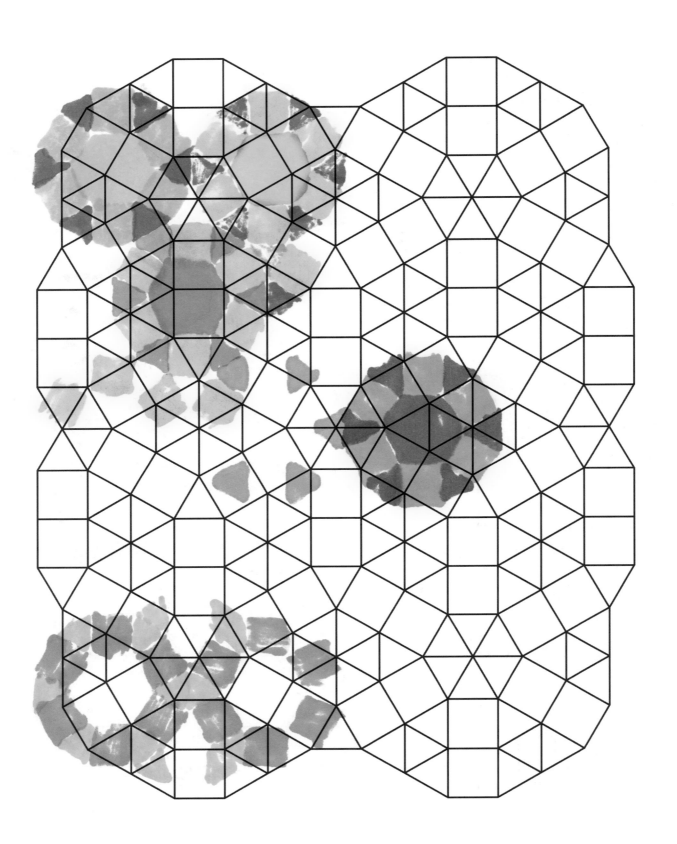

ACTIVITY 22 MASTER 22B Investigating Patterns/**Symmetry and Tessellations**

Islamic Tessellations

ACTIVITY SEQUENCE

MATERIALS

Master 23A, page 120 (handout)
Master 23B, page 121 (handout)
Colored markers

Discuss with students how artists, designers, and mathematicians have been interested in tessellations and their properties for centuries.

More than any other culture, Islam developed the art of geometric pattern, particularly in the years 700–1500. The early religious leaders of Islam interpreted Muhammad's preaching against idolatry as an injunction against the representation of humans or animals in art. Consequently, for centuries, Islamic art consisted of three types: designs derived from plant life, calligraphy, and repeating geometric shapes.

Hand out copies of Master 23A and 23B. Have students color each of the designs, repeating the colors of the shapes in a systematic way.

STUDENT CHALLENGE

Islamic tessellations are often based on a grid of squares or regular hexagons. The complex Islamic tessellation in Master 23A is based on a square, and uses only a compass and a straightedge to proceed from Step 1 to Step 4 below. Can you find the eight-pointed star of the tessellation in Step 4?

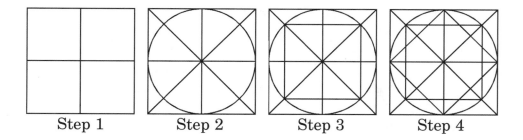

Step 1   Step 2   Step 3   Step 4

ACTIVITY 23 MASTER 23A   Investigating Patterns/**Symmetry and Tessellations**

The Art of M. C. Escher

ACTIVITY SEQUENCE

MATERIALS

Master 7C, page 40
    (transparency from
    Activity 7)
Master 24A, page 124
    (transparency)
Master 24B, page 125
    (transparency, 2 copies)
Master 24C, page 126
    (transparency, 2 copies)

VOCABULARY

parent polygon

Display the overhead of Master 24A as you present the following information to the class.

The master creator of tessellating shapes was Dutch graphic artist M. C. Escher (1898–1972). Escher's preoccupation with tessellations grew after a visit to the Alhambra in the 1930s. The walls, floors, and ceilings of this thirteenth-century fortress, built by Islamic Moors, are covered with mosaics of great variety and beauty. Escher spent days copying the patterns and remarked: "This is the richest source of inspiration that I have ever struck . . . What a pity it is that the religion of the Moors forbade them to make graven images!"

Although inspired by the Moorish mosaics, Escher preferred recognizable, animate figures to purely geometrical shapes. With extraordinary inventiveness, he created tessellating shapes that resemble birds, fish, lizards, dogs, humans, butterflies, and the occasional creature of his own invention.

Display the transparency of Master 7C, Escher's lithograph "Reptiles." Ask the following question.

1. *Look carefully! What else besides lizards do you see in the tessellation of lizards in Escher's sketchbook?*

    Escher included an underlying grid of hexagons. Observe that each reptile appears to fit into a hexagon in precisely the same way. Escher's creatures appear to be based upon tessellating polygons.

Display the transparency of Master 24B. Each bird in the tessellation is a modification of a specific tessellating polygon. To find this *parent polygon,* go around any of the interior birds with a pointer. Ask students to shout "stop" each time you reach a location or point where more than two tessellating shapes meet.

Pause briefly at each point. When you revisit your starting place, connect the "stop" points with your pointer. Ask students the following question.

2. *What kind of polygon did we outline by connecting points where more than two tessellating shapes meet?*

   a square

Display the transparency of Master 24C. Ask students if they can see how Escher modified the square to create the bird. The "bump" (sometimes called a "pimple") on the bottom of the square (the bird's legs) is identical (congruent) to the "hole" (sometimes called a "dimple") on the top of the square.

Align a duplicate of the transparency of Master 24C precisely over the first copy. Slide (translate) the overlay upwards until square vertices meet again.

The same is true for the left side and the right side of the square. The large bump and small hole on the left side are congruent to the large hole and small bump on the right side. Demonstrate with the overlay.

The tessellating bird shape has the same area as its parent square. For this tessellating shape, corresponding modifications are related by translation. However, the transformation or motion relating corresponding sides of a tessellating parent polygon will vary from tessellation to tessellation.

Align a duplicate of the transparency of Master 24B precisely over the first. Demonstrate how the corresponding tessellation has translational symmetry by sliding the overlay in various directions until the tessellating shapes coincide again. Include a slide where you align black birds with white birds. The effect will amaze most students.

LEARNING THE LANGUAGE

A *parent polygon* is a polygon that has been modified with congruent bumps and holes to create a non-polygonal tessellating shape.

STUDENT CHALLENGE

On the Internet, search for other artwork by M. C. Escher, particularly examples that involve mathematics. Escher was not a mathematician, but is greatly admired by them. Write a summary based on your findings.

ACTIVITY 24 MASTER 24A   Investigating Patterns/**Symmetry and Tessellations**

Investigating Patterns/**Symmetry and Tessellations**   ACTIVITY 24   MASTER 24B

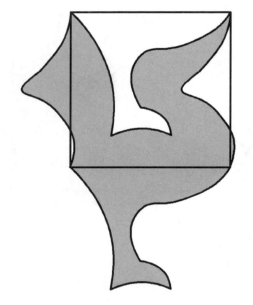

ACTIVITY 24 MASTER 24C Investigating Patterns/**Symmetry and Tessellations**

Making a Tessellating Template

ACTIVITY SEQUENCE

In this activity, students create their own artistic tessellating shape by modifying a square by translation. They will learn more rules for creating tessellating shapes in Activity 29.

Prepare cardboard squares by duplicating Master 25A on cardstock. The broken lines outline nine 2-inch squares. Each 2-inch square is divided into a 3 by 3 grid of smaller squares by the solid lines. Make as many copies as you need to give each student one 2-inch square. Allow for at least as many spares. Carefully cut along the broken lines with a fine utility knife and a steel ruler on a self-healing mat. Precision is essential!

Use scissors to extract the two irregular pieces from the remnants of the transparency of *Pieces of overlay* (used in Activity 15). Surround each piece with a narrow margin of clear acetate. The outlines of the pieces must be intact. Attach a small strip of reusable transparent tape to each piece along its straight edge, then tape the pieces in the appropriate holes of the lower-left figure on the transparency of Master 25B (see below).

MATERIALS

Master 25A, page 130 (copied on cardstock)

Master 25B, page 131 (transparency and a copy on cardstock)

Master 25C, page 132 (transparency)

Master 25D, page 133 (transparency)

Pieces of overlay, page 178 (transparency)

Utility knife

Steel ruler

Cutting mat

Reusable transparent tape

Scissors, 1 per student

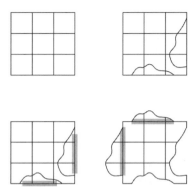

Prepare a tessellating template by copying Master 25B on cardstock. You can either cut out the figure in the lower-right corner, or you can cut out the square in top-right corner and prepare the template with scissors and tape as described below. The latter is recommended.

Align the vertices of a cardboard square with four dots in a square configuration on the transparency of Master 25C. Tell students they will be converting their square into a tessellating template using translation. In Activity 26, they will be using the dots on this handout to position their template when they draw their tessellation. Hence any changes they make to their square must not interrupt its corners.

Show the prepared transparency of Master 25B. The first figure depicts a cardboard square. Point out the simple "holes" drawn on each of two adjacent sides of the square of the top-right figure. Emphasize that the holes do not interrupt the corners of the square. Show students what happens if you remove the holes with scissors (detach the pieces in turn from the holes in the lower-left figure), then slide them to the opposite side (demonstrate on transparency). You know where to place the "bumps" by matching the grid lines. Clearly each hole must cross at least one grid line. To keep the "bumps" in place, use tape. The result is a non-polygonal tessellating template.

Lay the prepared cardboard template on the overhead. Turn it until the opaque projection (shadow) resembles a bunch of flowers with protruding stems. Can students visualize the interpretation? Keep turning it until you see a protruding nose. Tell students if they keep their shape simple, they can always turn the remnant between two holes into a silly nose.

Show the transparency of Master 25D. The student who designed this shape saw both an elephant and an elf. All it takes is a good imagination!

Allow students the time to create their own artistic tessellating template. Remind them to

keep their corners and keep it simple!

If students do not follow this advice, they may have problems in Activities 27 and 28. Besides, the simpler the outline of their tessellating shape, the easier it will be to give it a meaning. You may wish to check the students' holes before they cut them out.

As they complete their task, invite students to the overhead to interpret their shape by studying its projected contour. When they are done, tell students to add interpretive details to their template with a pencil.

STUDENT CHALLENGE

Make another tessellating template. Be creative and use your imagination!

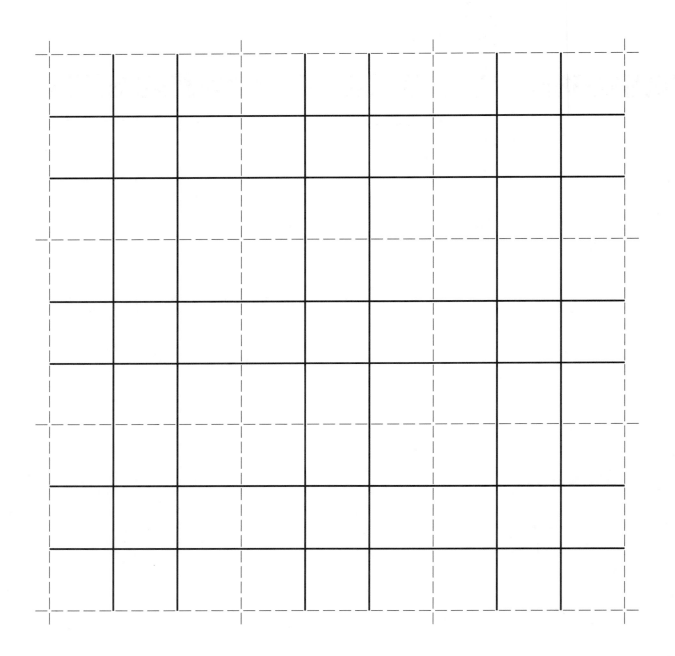

ACTIVITY 25 MASTER 25A Investigating Patterns/**Symmetry and Tessellations**

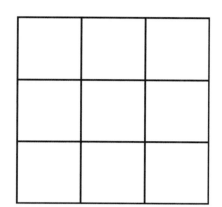

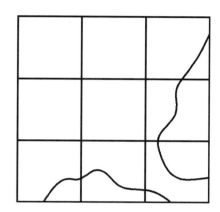

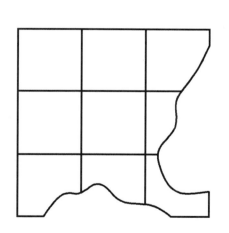

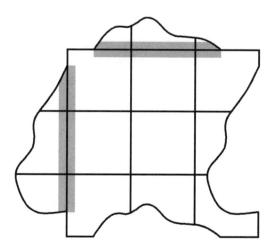

Investigating Patterns/**Symmetry and Tessellations**   ACTIVITY 25   MASTER 25B

ACTIVITY 25 MASTER 25C Investigating Patterns/**Symmetry and Tessellations**

Drawing a Tessellation with a Template

ACTIVITY SEQUENCE

MATERIALS

Tessellating templates from
    Activity 25 (cut out from
    Master 25B)
Master 25C, page 132
    (transparency and
    handout from Activity 25)
Master 26A, page 136
    (transparency)
Master 26B, page 137
    (transparency)
Master 26C, page 138
    (transparency)
Carbon paper, $\frac{1}{4}$ sheet per
    student (optional)
Ball-point pens, 1 per student
    (optional)

Do this activity immediately following Activity 25. Each student should already have a tessellating template. Give each student a copy of Master 25C.

Display the overhead transparency of Master 25C. Carefully align the vertices of your cardboard tessellating template from Activity 25 (cut out from Master 25B) with the four dots in the upper-left corner. Have students do the same with their own templates on their dot paper. Trace your template with a pointer and instruct students to trace tightly around their templates with sharp pencils.

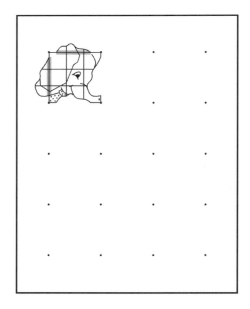

Remind students that they constructed their templates using translation, so they can continue their tessellation by sliding their templates to each new location. Demonstrate this on the transparency of Master 25C, then show the transparency of Master 26A.

Instruct students to trace only the undrawn edges of the shape. In other words, do not overlap markings. Thanks to the dots, the errors will be evenly distributed throughout the tessellation. Even the works of Escher himself had errors!

Show the transparency of Master 26B which includes interpretive details. How students add their features is up to them. As one method, students can draw the features freehand. Most students will be content with this approach.

Another method uses carbon paper to add precise features. Instruct students to place their templates faceup (grid lines showing) on a piece of carbon paper. Display an overhead of Master 26C and point to the top diagram. Have students follow these instructions:

- Trace the interpretive details of your shape with a pencil or ball-point pen. This deposits a layer of carbon on the back of the template. (Point to the bottom diagram on the transparency.)

- Go over each mark several times or press firmly as you trace to increase the carbon deposit.

- To draw the tessellation, trace around the template with a sharp pencil as before, then go over the interior marks as well. The carbon transfers to the dot paper.

- When you are done drawing the tessellation, redraw faint marks with your pencil.

STUDENT CHALLENGE

Using the template from the Student Challenge in Activity 25, create another tessellation.

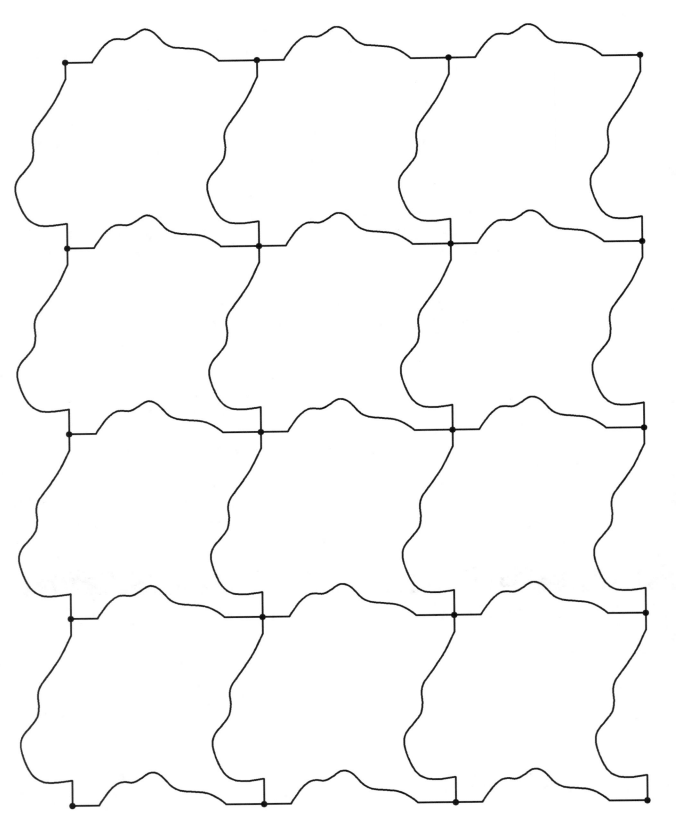

ACTIVITY 26 MASTER 26A Investigating Patterns/**Symmetry and Tessellations**

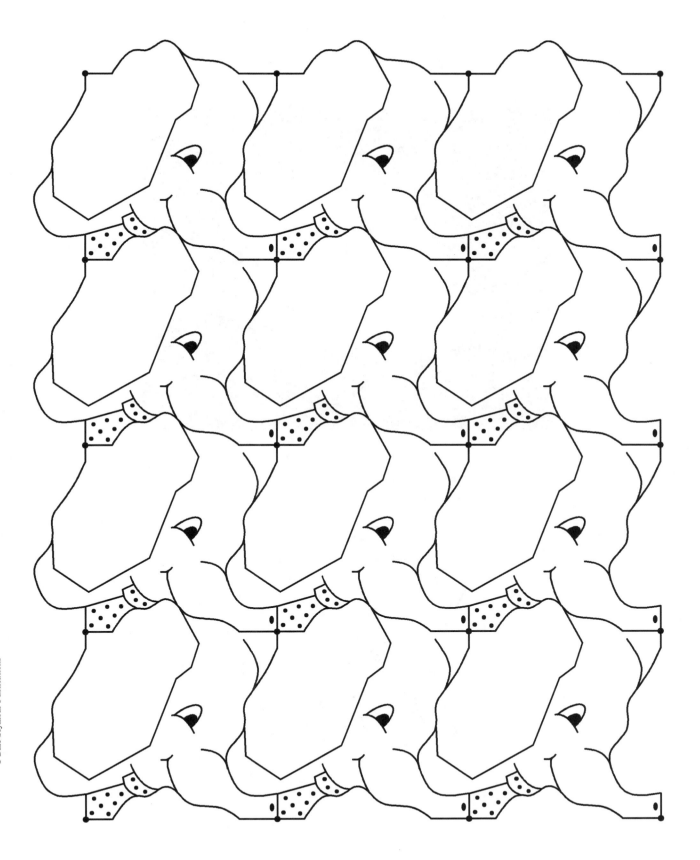

Investigating Patterns/**Symmetry and Tessellations** ACTIVITY 26 MASTER 26B 137

A Pop-Up Sponge
Jigsaw Puzzle

ACTIVITY SEQUENCE

MATERIALS

Master 27A, page 141
    (transparency)
Master 27B, page 142
    (transparency)
Pop-up sponges, $\frac{1}{6}$ full sheet
    per student
Student templates (created
    in Activity 25)
Resealable plastic bags, for
    storing dry sponge
Carbon paper, $\frac{1}{4}$ sheet per
    student (optional)
Black medium waterproof
    markers, about
    1 per pair
Containers of water

Classic jigsaw puzzles consist of interlocking pieces with different shapes that fit together in only one way. A tessellating shape could form the basis of a jigsaw puzzle with identical (congruent) pieces that fit together in many ways.

Preparing a compressed or "pop-up" sponge jigsaw puzzle is relatively expensive, but simply too much fun to pass up! Pop-up sponge comes in sheets that look like felt. You can draw on and cut it like cardboard. When put in water, the material expands or "pops" until approximately $\frac{1}{2}$-inch thick (a fascinating experience on its own!).

Incidentally, the dry material will swell when exposed to any source of moisture, including humid air, and will not return to its original thickness as it dehydrates. After opening the package, store the sheets in resealable bags. Dehydrated sponge will expand to the same thickness if immersed in water again.

Give each student two pieces of pop-up sponge ($\frac{1}{12}$ sheet of both white and blue). They should have their cardboard templates with carbon backing from Activity 25. Show students the transparency of Master 27A.

Have students trace their templates on each piece of sponge, drawing tightly around the template with a sharp pencil. The figure in the top-left corner of the transparency shows this method.

For the interior lines, students can either draw freehand or use carbon paper as described in Activity 26.

As an alternate method, students can trace the template on a piece of paper, add interpretive details, then transfer the entire drawing to the sponge with carbon paper and a pencil. (Point to the figure in the top-right corner of the transparency.)

Give students the following instructions for completing their tessellation puzzles.

• If you have not already done so, add interpretive details to each tracing.

• Retrace the features with a medium, black, waterproof marker (eliminating fine details). If you wish, you can highlight each shape's contour by going over the outside edges as well (point to the lower-left figure on the transparency).

• Immerse your shapes in water. (You may want to demonstrate using a sample sponge elephant.)

• Let the sponges completely expand, then squeeze out the excess water.

• Assemble your tessellating jigsaw puzzle.

I give each student four pieces of sponge (two pieces of each of two complementary colors). Each student makes a tessellating quartet. An elephant/elf tessellating sponge quartet is shown at the right.

You may decide to provide each student with a precut sponge of the elephant/elf region, then invite them to add their own interpretive details with waterproof markers. Your class assembly of the huge tessellating puzzle could look something like Master 27B (a sampling of selected interpretations that were created at the Escher Centennial Congress in 1998).

STUDENT CHALLENGE

Try tessellating sponge printing with stamp pad ink on dot paper. Print in a checkerboard configuration in one color, then fill in the spaces in a second color.

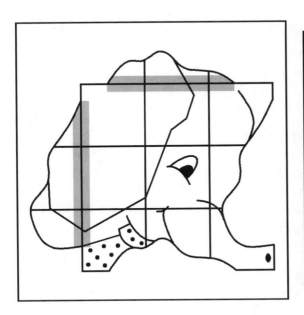

ACTIVITY 27 MASTER 27B Investigating Patterns/**Symmetry and Tessellations**

Mirroring the Craft of M. C. Escher

ACTIVITY SEQUENCE

This material was developed for a workshop that included George Escher, M. C. Escher's eldest son. George commented that he enjoyed the activity because "it mirrors what father did." Escher created lithographs and woodcuts. If you don't try printing, then you aren't mirroring Escher's craft.

Master 28A shows the steps to create a rubber stamp. The Appendix gives possible sources for the materials you will need.

Prepare the pieces of rubber by gluing plain paper to the waxy backing so students can draw on it, then cut it into 3-inch-by-4-inch pieces. [These dimensions allow for a larger "bump" in one direction.]

Pass out the materials. Students should have on hand their cardboard templates from Activity 25. Display the overhead of Master 28A. Give students the following instructions to make a rubber stamp.

- Lay your template on the paper backing of a piece of rubber.

- Trace tightly around the template with a sharp pencil or ball-point pen (Point to the figure in the upper-left corner of the transparency).

- Cut clockwise on the border of your shape, keeping the shape to the right of the scissors blades (point to the top-right figure). The paper and waxy backings will cut precisely, but the rubber tends to protrude beyond the border on the curves (center-left figure).

- Trim the rubber until it is flush with the boundary of the backing (center-right figure).

MATERIALS

Student templates (created in Activity 25)

Master 25B, page 131 (transparency from Activity 25, optional)

Master 28A, page 146 (transparency)

Master 28B, page 147 (transparency)

Master 28C, page 148 (transparency)

4-inch-by-4-inch transparent stamp mounts, 1 per student

3-inch-by-4-inch pieces of self-adhesive rubber, 1 per student

No-stick cooking spray

Scissors, 1 per student

Glue

Paper towels

Ball-point pens, 1 per student

11-inch-by-17-inch sheets of dot paper, 1 per student (master is provided in the envelope in the back of this book)

Newspaper, several sheets per student

Large stamp pads ($4\frac{1}{2}$ inch by $7\frac{1}{2}$ inch), 1 per 4 students in at least 2 colors

Freezer bags, about 6, at least 5 inches square

Rubbing alcohol

- Oil the flat plate of the mount with no-stick cooking spray.

- Peel the backing from the rubber shape to expose its self-adhesive surface. Attach the rubber to the mount (lower-left figure), and wipe off any excess oil with a paper towel.

- Add interior interpreting features by scoring the surface of the rubber using short repetitive strokes with the pencil or a ball-point pen (lower-right figure).

Give each student a sheet of 11-inch-by-17-inch dot paper and several sheets of newspaper. Students should lay the dot paper on the newspaper.

Cut a 3-inch diameter circle in the center of about six freezer bags. Lay a freezer bag over each stamp pad, centering the hole on the pad. The bags are "shields" to prevent inking the mount along with the rubber stamp. No more than four students sharing one stamp pad works best.

Give students the following instructions for printing their tessellations.

- Saturate the rubber printing shape with ink by pressing the stamp firmly into the stamp pad. Make sure that your complete rubber shape falls within the hole in the shield and that the **entire** plate of the mount contacts the pad. Adjust the location of the shield as required.

- Align the vertices of the parent square of the rubber shape over one set of dots on the dot paper. (You may demonstrate with an elephant stamp on a transparency of dot paper such as Master 25C).

- Lower the mount down, vertex to corresponding dot. Apply a firm, even pressure to the top of the mount, then lift straight up to expose the colored impression of your shape.

- Fill the dot paper with imprints in a checkerboard pattern, reinking the stamp each time. (Show the transparency of Master 28B.)

- When done, clean the stamp under running water, dry it, then repeat in the spaces using a contrasting color of stamp pad ink. (Show the transparency of Master 28C.)

- When your print is complete, clean the stamp once again, then tear the rubber region off the mount. Remove any leftover adhesive on the mount with rubbing alcohol.

STUDENT CHALLENGE

Prepare a second stamp of the same tessellating template, then add a completely different set of interpretive details. Print in a checkerboard pattern once again, using a different stamp for each color (see below).

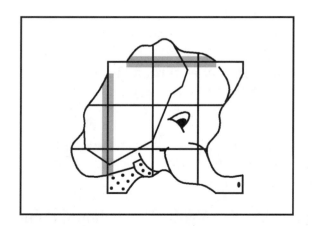 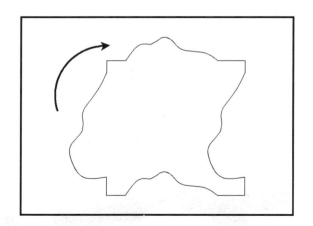

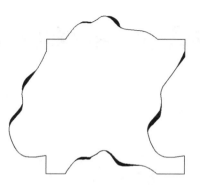 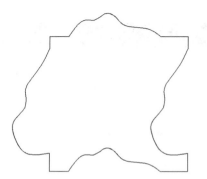

 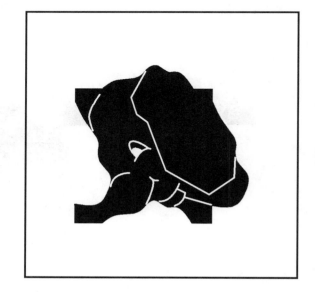

ACTIVITY 28 MASTER 28A   Investigating Patterns/**Symmetry and Tessellations**

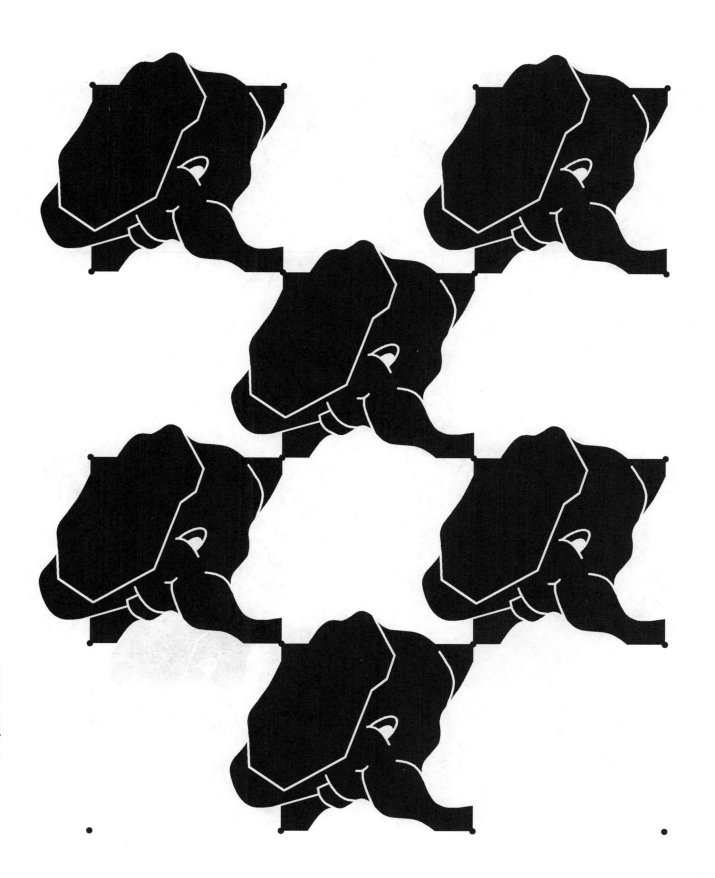

ACTIVITY 28 MASTER 28C Investigating Patterns/**Symmetry and Tessellations**

Exploring
Tessellating Art

MATERIALS

Masters 29A, 29D, 29H, 29K, 29N, 29Q, and 29T, pages 154, 157, 161, 164, 167, 170, 173 (transparency, duplicate optional)

Masters 29B, 29E, 29I, 29L, 29O, 29R, and 29U, pages 155, 158, 162, 165, 168, 171, 174 (transparency and handout)

Masters 29C, 29F, 29G, 29J, 29M 29P, and 29S, pages 156, 159, 160, 163, 166, 169, 172 (transparency)

Master 29V, page 175 (transparency and handout copied on translucent drawing paper, sample provided in envelope in the back of this book)

Overhead marker

Rulers, 1 per student

Sheets of dot paper, 1 per student (sample provided in envelope in the back of this book)

ACTIVITY SEQUENCE

Now that students have some experience creating tessellating art, they are ready to learn more ways to create these symmetrical patterns.

Explain to students that they will be shown seven of M. C. Escher's tessellations. Each will be followed by a corresponding worksheet. They will use each worksheet twice—once to reveal the parent polygon and once to discover the modifying rule or set of rules. When students made artistic tessellations previously, the parent polygon was a square and the modifying rule was translation. They slid a bump (or hole) on one side of a square to the side directly opposite it, forming an identical hole (or bump). The area of the square did not change. Now students will encounter other modifying rules.

Show the transparency of Master 29A, asking students to study the Escher tessellation. Provide each student with a copy of Master 29B, the corresponding worksheet. Review how the parent polygon was revealed in Activity 24. Ask them to go around the border of the shaded shape on their worksheet, looking for locations (points) at which more than two tessellating shapes meet. If they connect all such points in a circular (cyclic) order, they will reveal the parent polygon. (It is a square.) Have students draw this square on their worksheet using a ruler and pencil. Demonstrate by adding the square to the transparency of the worksheet.

Repeat this process for Masters 29D (Escher tessellation) and 29E (corresponding worksheet). Show the Escher tessellation, have students reveal the parent polygon using their worksheet, then have them draw the parent polygon on their worksheet as you demonstrate at the overhead. (It is a parallelogram.) Repeat with Masters 29H and 29I (another square); Masters 29K and 29L (a

ACTIVITY 29 • Exploring Tessellating Art 149

hexagon); Masters 29N and 29O (an equilateral triangle); Masters 29Q and 29R (another parallelogram); and finally Masters 29T and 29U (a quadrilateral kite). Point out in Master 29Q (Escher's ingenious bulldogs) how the tessellation suffers from "foot in mouth" disease.

Give each student Master 29V, the solutions, copied on translucent drawing paper. The envelope in the back of this book contains a sample. The Appendix gives specifications for the material and possible sources. Ask students to correct their worksheets using this solution sheet.

Have students look for corresponding bumps and holes on the sides of each parent polygon on the solution sheet. A simple visual inspection will suffice. If you wish, direct these preliminary investigations by using a transparency of Master 29V to point out a few identical (congruent) bumps and holes.

When satisfied, ask students to overlay and align the appropriate shaded region on the solution sheet, in turn, with the shaded region on each worksheet. Consider the worksheets in the order of their initial presentation. Ask students to propose a modifying rule (or a combination of modifying rules) for each tessellating shape. In order to help them grasp these concepts they will need to ask themselves how Escher made each tessellation. It is imperative for students discover these on their own.

Everyone tests a proposed rule by attempting to use it to move the shape on the translucent overlay so that it aligns precisely with an unshaded adjacent shape on the worksheet. (By an adjacent shape I mean one that shares more than a single point with the shaded shape on the worksheet.) They will be able to do this in four potential ways. [Note: Three of the examples (Master 29E, Master 29O, and Master 29R) will involve a combination of two distinct modifying rules.]

Translation

Modifications translate between a pair of equal and parallel sides of the parent polygon. Bumps/holes on one side become congruent holes/bumps on the equal and parallel opposite side. You can slide an overlay of the shape to coincide with an adjacent shape (shares more than a point) in the tessellation.

Midpoint Rotation

Modifications rotate 180° around the midpoint of a side of the parent polygon. Bumps/holes on one half-side become congruent holes/bumps on the other half-side. You can turn an overlay of the shape 180° around the midpoint to coincide with an adjacent shape in the tessellation.

Vertex Rotation

Modifications rotate about a vertex of the parent polygon between a pair of equal and adjacent sides. A hole/bump on one side becomes a congruent bump/hole on an equal and adjacent side. You can turn an overlay of the shape about the vertex to coincide with an adjacent shape in the tessellation. The common vertex acts like a pivot point or hinge.

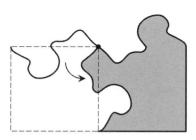

Glide Reflection

Modifications glide reflect (flip and slide) between a pair of equal sides of the parent polygon. You can flip an overlay of the shape over, then slide it until it coincides with an adjacent shape in the tessellation.

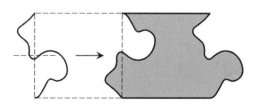

Again I must stress that these rules should not be read to the class. Allow students to find them through their investigations. Point out that one side of the parent polygon on the translucent overlay must coincide exactly with a side of the parent polygon on the worksheet following any valid move. Each time a student proposes a rule, the class will either verify it or reject it. Direct the activity using the transparency of the solution sheet on the transparencies of the worksheets. As these examples are completed, show the class the transparencies of the reinforcing examples provided. These additional examples feature a shaded shape and the parent polygon. The solutions follow.

Master 29B uses translation between each pair of parallel and equal sides of the parent square. The reinforcing example (Master 29C), a tessellation of sleepy buzzards, uses the same rule, however the parent polygon is a parallelogram instead of a square. Rectangles or hexagons with three sets of parallel and equal sides (see Master 18E) can be modified in the same way.

Master 29E uses two different rules, namely translation between one pair of parallel and equal sides of the parent parallelogram, and midpoint rotation on each of the other two sides. Both this tessellation and the tessellation of buzzards have a parallelogram as a parent polygon, however their modifying rules differ. The first reinforcing example (Master 29F), a dog with a bow tie, has a rectangle as a parent polygon. The second (Master 29G), a tessellation of parrots, used midpoint rotation on all four sides of a parent scalene quadrilateral. (In Activity 18 you found that any quadrilateral will tessellate by midpoint rotation.)

Master 29I uses vertex rotation between two pairs of equal and adjacent sides of the parent square. Two opposite vertices of the square, specifically the ones between the related sides, are centers of rotation. The angle of rotation at each of these vertices is 90°. The reinforcing example (Master 29J), a charming tessellation of bunnies, is based on the same modifying rule.

Master 29L uses vertex rotation between three pairs of equal and adjacent sides of the parent hexagon. The hexagon is irregular. However it does have three 120° angles, and the arms of these angles are equal in length (see Activity 18). Every second vertex is a center of rotation, and each angle of rotation is 120°. The reinforcing example (Example 29M), a battle between peg-legged pirates, uses the same modifying rule on a regular hexagon.

Master 29O uses vertex rotation between one pair of equal and adjacent sides as well as midpoint rotation on the third side of the parent equilateral triangle. The angle of rotation for the vertex rotation is 60°, as would be expected for an equilateral triangle. The reinforcing example (Master 29P) uses the same set of modifying rules.

Master 29R uses translation between one pair of equal and parallel sides of the parent parallelogram, and vertical glide reflection between the other pair. The reinforcing example of goldfish (Master 29S) is identical in construction.

Master 29U uses vertical glide reflection between each pair of equal sides of the parent quadrilateral kite. The configuration of tessellating horsemen is similar to the configuration of kites in the corresponding polygonal tessellation (see Activity 18), but without the horizontal reflectional symmetry.

Following these activities, you may wish to revisit the Escher tessellations. Use a duplicate of each transparency as an overlay on the corresponding original transparency to demonstrate its symmetries. All have translational symmetry. The last two have glide-reflectional symmetry. The others have some form of rotational symmetry. If midpoint rotation is involved, the order of rotation is two. If vertex rotation is involved, the order will be two, three, four, or six. For each tessellation, ask students to provide the corresponding angle of rotation: 180°, 120°, 90°, or 60°.

Several of the Escher tessellations that involve midpoint or vertex rotation have more centers of rotation than their modifying rule or rules would lead us to expect. Master 29D has centers of order two at the right elbow, the left knee, the tip of the left hand, and the tip of the right foot of each reptile (in addition to the anticipated centers of order two owing to the midpoint rotation). Master 29H has centers of order two at both the nose and right knee of each reptile (in addition to the centers of order four owing to the vertex rotation of the parent square). Master 29N has centers of order three at the tip of the left wing and the tip of the tail of each bird (in addition to the centers of order two and six owing to the midpoint rotation and vertex rotation respectively).

All this can lead an Escher novice to believe that the geometrical basis of the artist's tessellations is more complex than in actuality. In any event, it is Escher's ingenious outlines, that are most admired—even by mathematicians!

STUDENT CHALLENGE

A reinforcing example was not provided for the last of the seven Escher tessellations, his famous horsemen. Try to create your own. Use a cardboard template of the parent quadrilateral kite and appropriate dot paper to simplify your task.

ACTIVITY 29 MASTER 29A Investigating Patterns/**Symmetry and Tessellations**

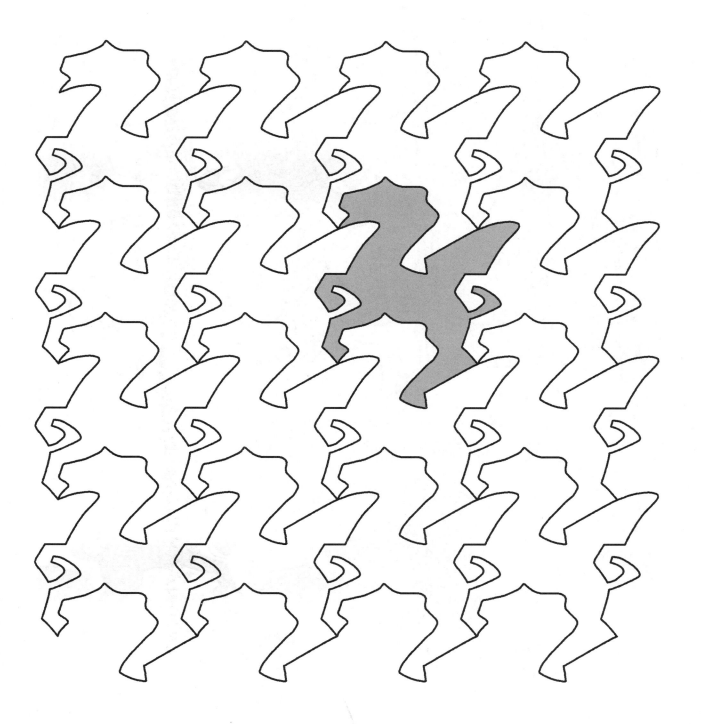

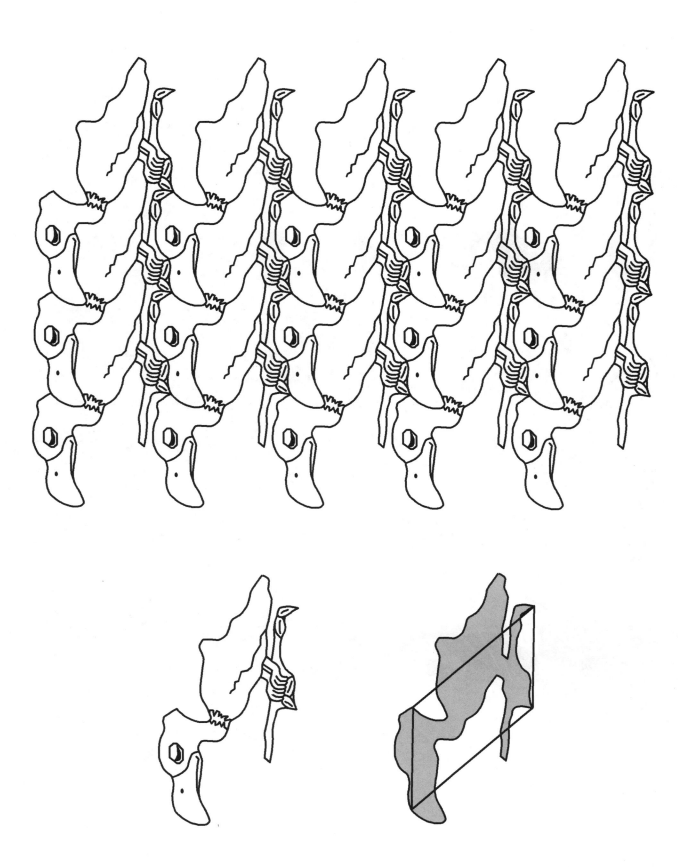

ACTIVITY 29 MASTER 29C Investigating Patterns/**Symmetry and Tessellations**

ACTIVITY 29 MASTER 29E Investigating Patterns/**Symmetry and Tessellations**

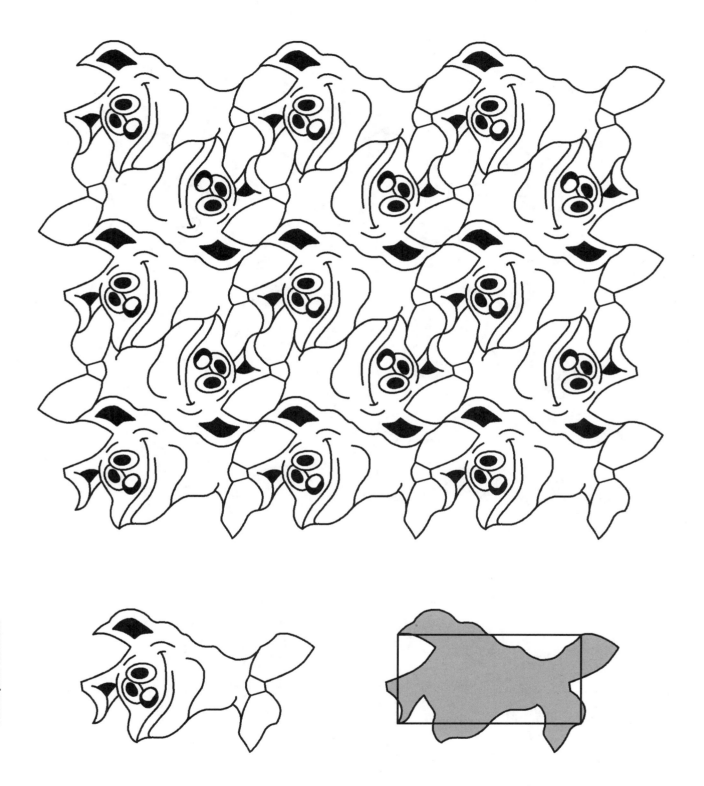

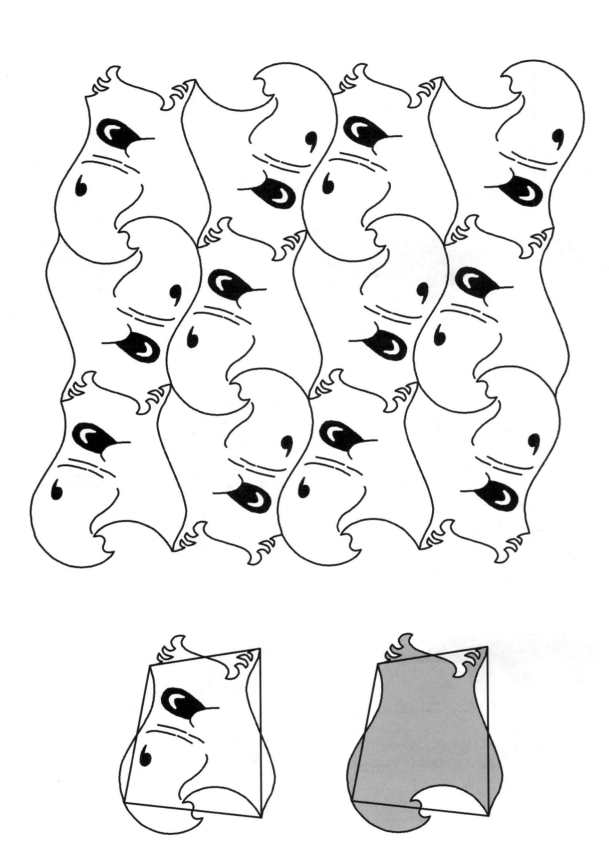

ACTIVITY 29 MASTER 29G Investigating Patterns/**Symmetry and Tessellations**

ACTIVITY 29 MASTER 29I Investigating Patterns/**Symmetry and Tessellations**

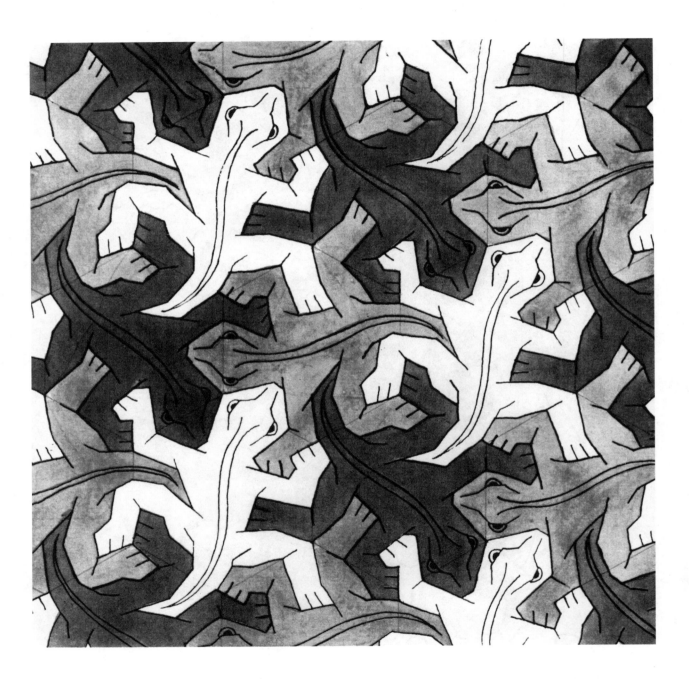

ACTIVITY 29 MASTER 29K Investigating Patterns/**Symmetry and Tessellations**

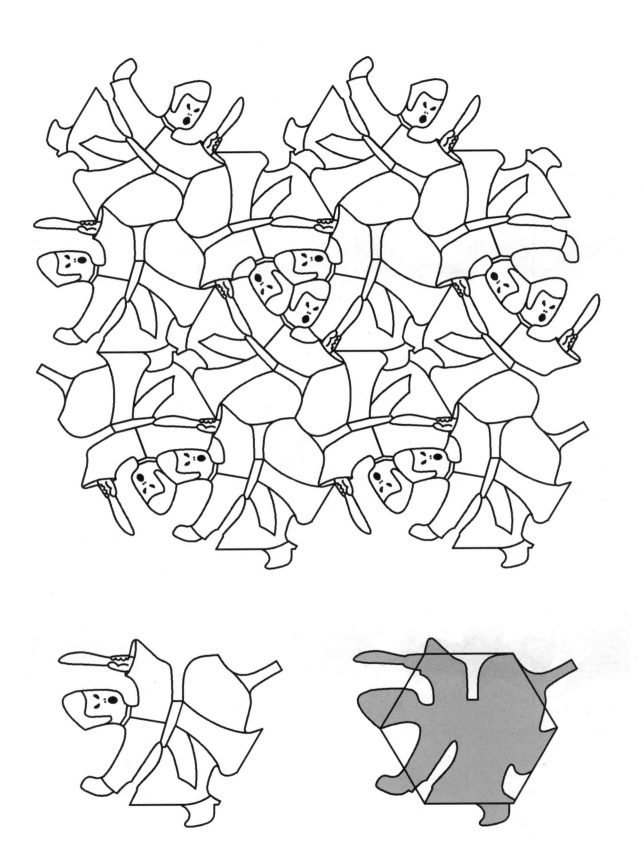

ACTIVITY 29 MASTER 29M   Investigating Patterns/**Symmetry and Tessellations**

ACTIVITY 29 MASTER 29O Investigating Patterns/**Symmetry and Tessellations**

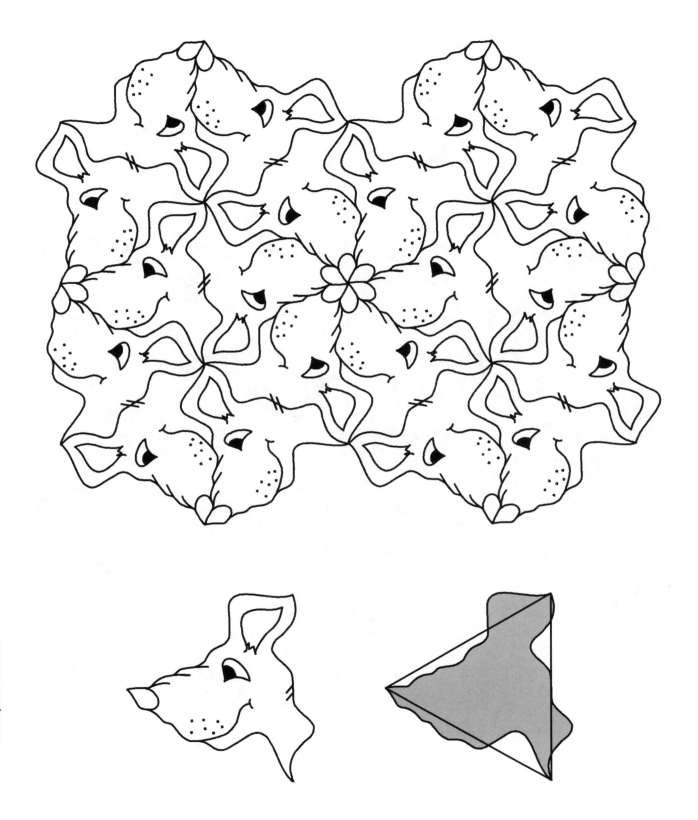

ACTIVITY 29 MASTER 29Q Investigating Patterns/**Symmetry and Tessellations**

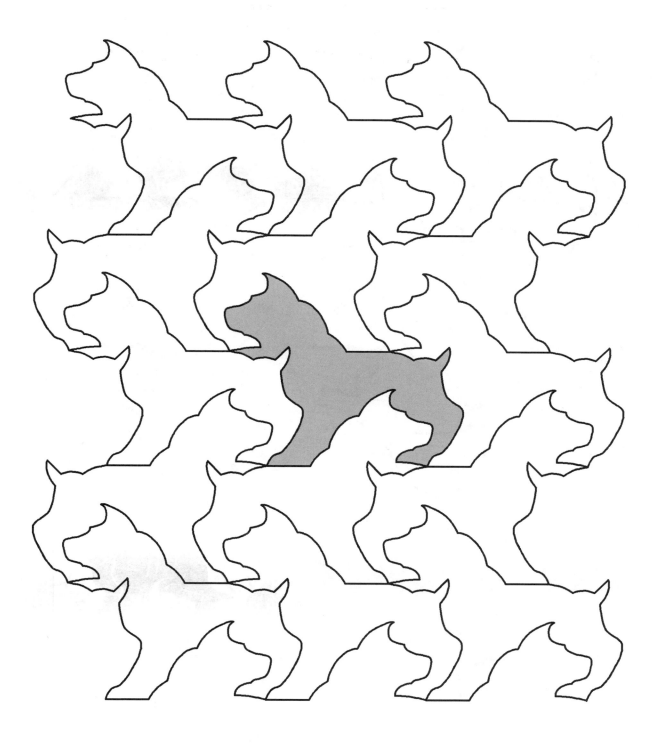

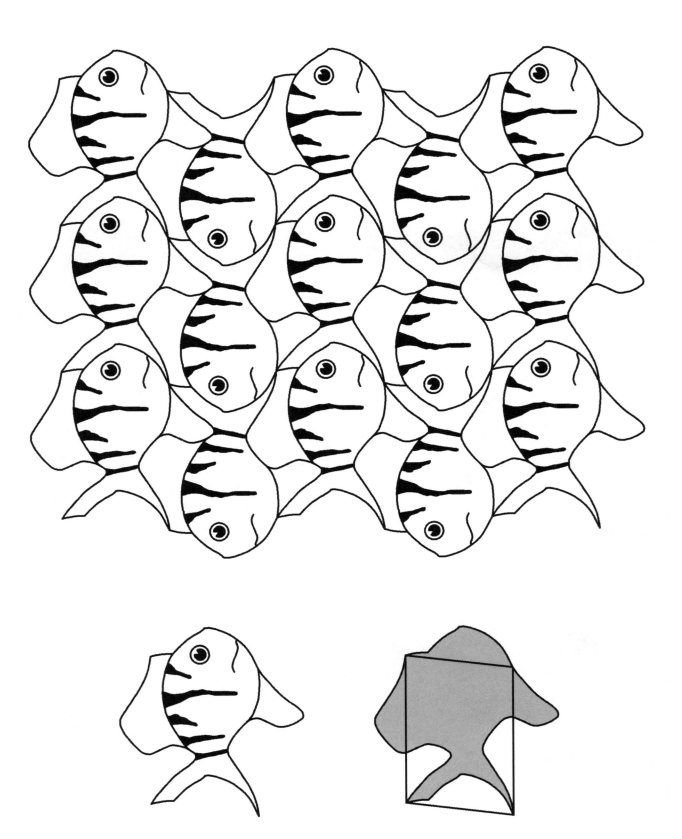

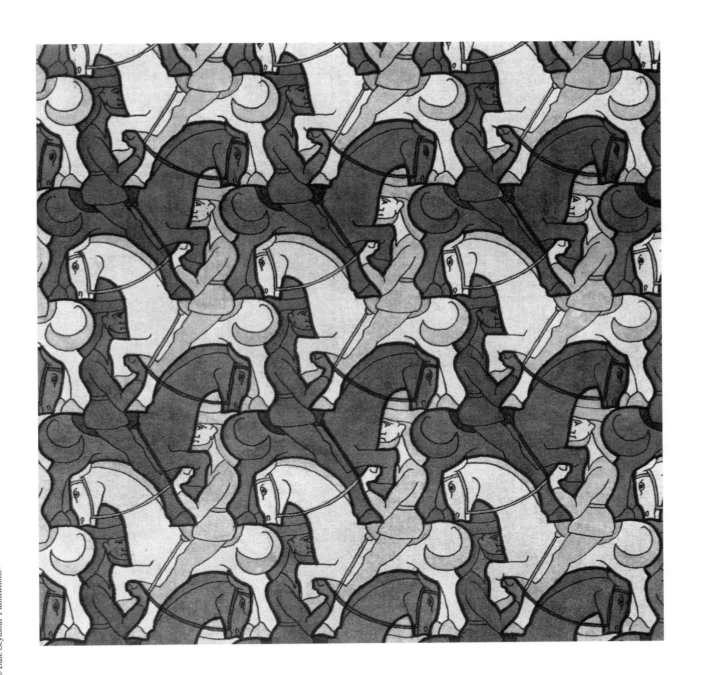

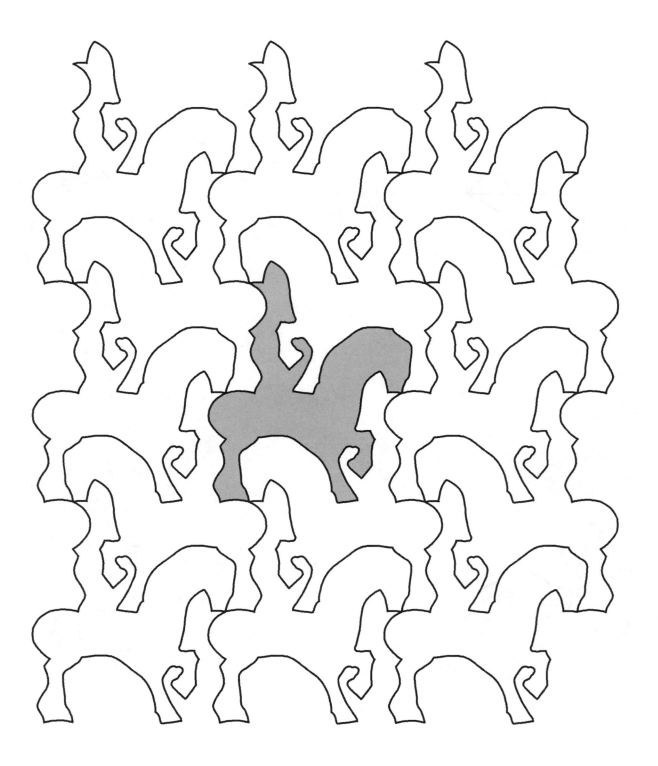

ACTIVITY 29 MASTER 29U   Investigating Patterns/**Symmetry and Tessellations**

Explorations with *TesselMania!*®

ACTIVITY SEQUENCE

MATERIALS

TesselMania!® computer program
(demo or full version)
T-shirts, I per student (optional)
Iron-on T-shirt transfers
(optional)

In this closing activity, students create tessellating art on a computer with *TesselMania!*® software, created by Kevin Lee. The program's simple interface makes it possible for them to produce and investigate intricate tessellations quickly. You can download a demo version of *TesselMania!*® (Windows or Macintosh) from the following internet site: http://worldofescher/store/tessftp.html.

Read through the instructions (under the Help command) to learn how to use *TesselMania!*®. The program first asks students to choose a modifying rule or set of rules (translation, rotation, and/or glide reflection). Then students choose a tessellating tile (a triangle, a quadrilateral, or a hexagon) to which the rule or rules will be applied. The drawing screen includes a tack tool to add bumps or holes to any side of the tessellating tile. The program automatically removes or adds the corresponding holes or bumps from or to the appropriate related side according to the rule or rules selected.

Classic paint tools, including stamps, are available for adding interior features. When their modified tile is complete, students simply click on the tessellate button, and the corresponding tessellation fills the screen automatically. Other buttons allow students to animate the construction of their modified tile from its parent polygon, to simulate the drawing of their tessellating artwork, and to animate the metamorphosis of their artwork to/from its parent tessellation. Escher would have loved it!

Have each student create an example of tessellating artwork using the software. The demo works exactly like the full version of *TesselMania!*®, but the save and print options have been disabled.

If you have the full *TesselMania!®* program, you can make T-shirts of the students' creations. Simply print the artwork on T-shirt transfers using a color bubblejet printer, then iron the transfers onto T-shirts. The results are outstanding!

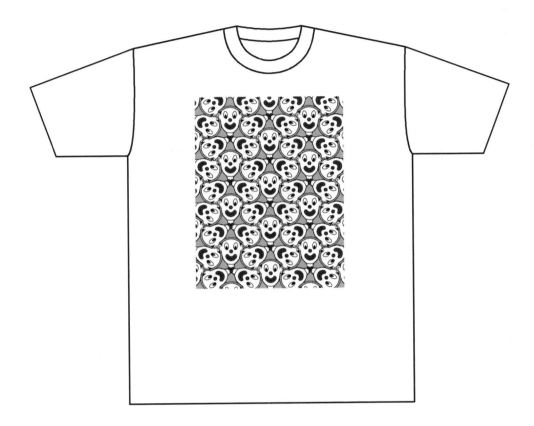

STUDENT CHALLENGE

Now that you have experience with tessellations and symmetry, create your own student challenge for other students to solve.

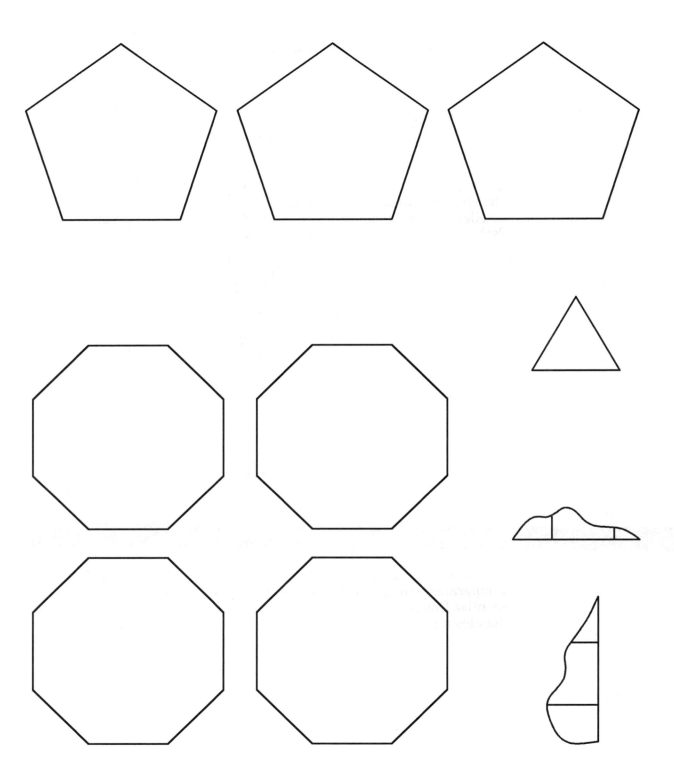

Pieces of overlay   Investigating Patterns/**Symmetry and Tessellations**

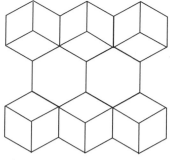

APPENDIX

The following gives additional information for each activity, including sources for materials, extension activities, and related Internet sites.

Activity 1
Showing the Disney video *Donald in Mathmagic Land* provides excellent reinforcement to this activity. Donald Duck meets Pythagoras, then encounters mathematics in music, architecture, art, games, and a myriad of everyday things.

The following Internet site details Pythagoras' observations on the relationship between music and number.
   http://www.aboutscotland.co.uk/harmony/prop.html

Activity 2
This Internet site has galleries containing butterfly and moth images. In most cases, a larger picture is available by clicking on the small picture.
   http://mgfx.com/butterfly/gallery/index.htm

This Internet site contains graphics of the most commonly used traffic signs.
   http://member.aol.com/rcmoeur/signman.html

This Internet site has a trial version of a symmetric art program.
   http://users.penn.com/~hank/symtoy.html

Activity 3
An "Alphabet Symmetry" poster and a set of "Inversions" posters, both by Scott Kim, are available from Dale Seymour Publications. Scott maintains an Internet site with several of his latest inversions (see net address below), some of which are animated. You will also find other "ambigrams" (the generic term for inversions, coined by cognitive scientist Douglas Hofstadter) on the web. A site by David Holst includes an automatic ambigram generator that produces ambigrams with rotational symmetry for any word or phrase you type in.

This Internet site features several of Scott Kim's inversions.
   http://www.scottkim.com/inversions/

This Internet site has an ambigram generator.
   http://ambigram.matic.com/ambigram.htm

Activity 4
The It's Okay If You Sit on My Quilt Book by Mary Ellen Hopkins, available from ME Publications, contains 357 color examples of geometric quilt blocks on a gridded background. It amounts to an

encyclopedia of quilt block designs for the geometrically-inclined. A must have book!

This Internet site features national and signal flags.
http://155.187.10.12/flags/signal-flags.html

This Internet site features quilt block patterns.
http://www.portup.com/~hjbe/quilt/

Activity 5
Car wheel covers often show rotational symmetry. A poster on "Symmetry in Wheels" is available from Dale Seymour Publications. You will also find wheel covers at the Internet address below. A colorful poster of hex signs is available from the Folkart & Craft Exchange (see net address below).

This home page of Style-Line features symmetrical wheel covers.
http://style-line.com/

This Internet site from the Folkart & Craft Exchange features hex signs and Pennsylvania hex signs.
http://www.folkart.com/~latitude/hex/hexx.htm

Activity 6
There are seven unique combinations of symmetries that can be found in a strip pattern or frieze. The first Internet site below shows each of the seven types and provides simple names for each. The second site contains examples of all seven types in cast iron decorative work.
http://www.geom.umn.edu:80/~lori/kali/friezegrp.html
http://d-newton.educ.cam.ac.uk/mathsf/journalf/nov98/friezes.html

Activity 7

Through this Internet site, you can purchase the *Lizards of M.C. Escher SoftPuzzle*, an officially licensed adaptation of Escher's famous reptile. Each contains 15 soft foam pieces that can be arranged in many different ways, but always tile together perfectly. The completed puzzle is the size of a small desktop. It provides an excellent introduction to tessellations once students complete the connect-the-dots puzzle.
http://www.iproject.com

Activity 8
In this Internet site, Dr. Math provides prefixes for the first 50 polygons, among others.
http://forum.swarthmore.edu/dr.math/faq/faq.polygon.names.html

This Internet site features animated dissections of polygons into other polygons.
http://www.cs.purdue.edu/homes/gnf/book/webdiss.html

This Internet site features a Snow Crystal Display, a small but unique collection of Wilson A. Bentley's snowflake photomicrographs.
http://www.snowflakebentley.com/index.htm

Activities 10, 11, and 12
I owe my inspiration for these activities to Linda Silvey and Loretta Taylor's informative book *Paper and Scissors Polygons and More,* available from Dale Seymour Publications. In this book, you can find

proof that the folds of a paper and scissors pentagon and hexagon will create the required angle.

Kirigami is the Japanese art of folding and cutting paper. This Internet site lets you create similar figures by cutting a wedge from a virtual piece of folded paper.

http://www.thegrid.net/kevinsw/web/kirigami.html

This is the ultimate Origami (paper-folding) site featuring hundreds of links.

http://ccwf.cc.utexas.edu/~vbeatty/origami/

Activity 15

The first Internet site below will get you "Totally Tessellated." The second site has thumbnails of all graphics in the first (including 17 Escher tessellations). Click on an image to open a new window with a larger image and more.

http://library.advanced.org/16661/
http://library.advanced.org/16661gallery/thumbnails.html

This Internet site contains a link to a Patterns Program in which you can drag pattern block shapes into a working area where they can be translated and rotated.

http://www.forum.swarthmore.edu/sum95/suzanne/active.html

Activity 16

This Internet site from the San Francisco Exploratorium has information on bubbles, and includes links to other sites on bubbles as well as a "bubbliography."

http://www.exploratorium.edu/ronh/bubbles/

Activity 17

You can investigate and generate symmetric patterns with *KaleidoMania! Interactive Symmetry* (by *TesselMania!* creator Kevin Lee), available from Key Curriculum Press. This captivating new program includes several activities to teach the mathematics of symmetry along with a kaleidoscopic-like screen that allows the student to generate symmetric patterns using their own images and designs. A poster on "Kaleidoscope Symmetry" is available from Dale Seymour Publications.

This Internet site from The Kaleidoscope Collector shows how a kaleidoscope works.

http://www.kaleidoscopesusa.com/how.htm

This Internet site features an interactive and ever-changing Java Kaleidoscope.

http://www.michaels.com/main/kaleidoscope/kscope-gam.html

Activity 20

A poster of "Tessellations" and a set of 12-inch-by-12-inch translucent "Stained Glass Tessellations," whose patterns represent the three regular and eight semiregular tessellations, are available from Dale Seymour Publications.

Activity 22
Further examples of demiregular tessellations are found on page 57 of *Introduction to Tessellations* by Dale Seymour and Jill Britton, available from Dale Seymour Publications. On pages 58 and 59 of the same book, you will find examples of tessellations of regular polygons in which the polygons are not arranged side to side and vertex to vertex.

This "Totally Tessellated" Internet site allows you to print full-page renditions of regular, semiregular, and demiregular tessellations suitable for coloring activites.
http://hyperion.advanced.org/16661/templates/

Activity 23
You can recreate Islamic art with *TesselMania!*® software (see Activity 30). Suitable activities are found in *Explorations with TesselMania!*® by Jill Britton and are available from Dale Seymour Publications. The book includes a disk of files generated with *TesselMania!*®, including several of the Islamic type.

An excellent treatise on Islamic designs is found on pages 172–179 of *Introduction to Tessellations* by Dale Seymour and Jill Britton, available from Dale Seymour Publications.

The Metropolitan Museum of Art, New York, has prepared *The Mathematics of Islamic Art,* and refers to it as "a packet for teachers of mathematics, social studies, and art." It includes 16 slides, some featuring objects from the museum's collection of Islamic art. Suggestions for using the materials with middle school students are provided as well.

This Internet site is called Symmetry and Pattern: The Art of Oriental Carpets.
http://forum.swarthmore.edu/geometry/rugs/

Activity 24
I highly recommend showing the 22-minute film *Adventures in Perception,* which details the life and graphic works of Escher. Produced in Netherlands two years before Escher's death, it includes a clip of the artist painstakingly carving his last woodcut, *Snakes,* then printing the graphic.

The video *The Fantastic World of M. C. Escher* is available from Dale Seymour Publications, as well as Escher books, posters, and the CD for Windows *Escher Interactive.* The *M. C. Escher Coloring Book* is a wonderful source of line drawings.

These Internet sites each contain several of Escher's graphic works, including his tessellations.
http://www.nga.gov/collection/gallery/ggescher/ggescher-main1.html
http://www.worldofescher.com/
http://www.nga.gov/cgi-bin/psearch?Request=A&Person=201590
http://www.mathacademy.com/platonic_realms/minitext/escher.html
http://hyperion.advanced.org/11750/

Activity 26

More ideas for making a tessellating template and what to do with it are found in *Teaching Tessellating Art* by Jill Britton and Walter Britton, available from Dale Seymour Publications.

Activity 27

Pop-up sponge is available through Dale Seymour Publications in packages of six 9-inch-by-11-inch sheets. Three sheets are blue, three sheets are white.

Activity 28

Dale Seymour Publications carries 4-inch-by-4-inch transparent stamp mounts. The mounts are reusable and come in sets of four. Also available from Dale Seymour Publications are 9-inch-by-12-inch sheets of self-adhesive rubber. You must glue plain paper to the waxy backing of this rubber so students can draw on it.

Activity 29

I recommend using Staedtler© Mars-Vellum High Quality Drawing Paper (No. 946 811S) for the translucent copy of the solutions on Master 29V. It is available in packages of 50 sheets at office supply stores.

This Internet site has links to tessellations around the world—including animated ones.
     http://dspace.dial.pipex.com/crompton/Home.shtml

Activity 30

TesselMania!® and the more elaborate CD-version *TesselMania!*® *Deluxe* are available through Dale Seymour Publications.

T-shirt transfers for most color bubblejet printers are available from computer supply stores.

In my book *Explorations with TesselMania!*®, also available from Dale Seymour Publications, I show you how to use the software to recreate tessellations, including those of M. C. Escher, to reproduce quilt blocks, and to recreate several classic Islamic art designs. The accompanying disks contain numerous examples of original student art, as well as files to help generate all of the above.

This Internet site allows you to download a demo version of *TesselMania!*® (Windows or Macintosh).
     http://www.worldofescher.com/store/tessftp.html

**For additional information regarding products by
Dale Seymour Publications and for links
to Jill Britton's site, visit this website:
http://www.cuisenaire-dsp.com**

**For direct access to Jill Britton's page of links
coordinated with the activities in this book,
visit this website (updated periodically):
http://www.camosun.bc.ca/~jbritton/jbsymteslk.htm**

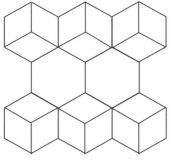

BIBLIOGRAPHY

Bezuszka, S., M. Kenney, and L. Silvey. *Tessellations: The Geometry of Patterns.* Palo Alto, CA: Creative Publications, 1977.

Bool, F. H., J. R. Kist, J. L. Lochner, and F. Wierda. *M. C. Escher: His Life and Complete Graphic Work.* New York: Harry N. Abrams, Inc., 1981.

Britton, J., and W. Britton. *Teaching Tessellating Art.* Palo Alto, CA: Dale Seymour Publications, 1992.

Britton, J. *Explorations with TesselMania!®* Palo Alto, CA: Dale Seymour Publications, 1997.

Ernst, B. *The Magic Mirror of M. C. Escher.* Norfolk, England: Tarquin Publications, 1985.

Escher, M. C. *The Graphic Work of M. C. Escher.* New York: Ballantine Books, 1967.

Hopkins, M. E. *The It's Okay If You Sit on My Quilt Book.* Santa Monica, CA: ME Publications, 1989.

Kennedy, J., and D. Thomas. *Kaleidoscope Math.* Palo Alto, CA: Creative Publications, 1978.

Lappan, G. *Kaleidoscopes, Hubcaps, and Mirrors.* Palo Alto, CA: Dale Seymour Publications, 1998.

Lochner, J. L., ed. *The World of M. C. Escher.* New York: Harry N. Abrams, Inc., 1971.

MacGillavry, C. H. *Fantasy & Symmetry: The Periodic Drawings of M. C. Escher.* New York: Harry N. Abrams, Inc., 1976.

Schattschneider, D. *Visions of Symmetry: Notebooks, Periodic Drawings, and Related Work of M. C. Escher.* New York: W. H. Freeman, 1990.

Seymour, D. *Tessellating Teaching Masters.* Palo Alto, CA: Dale Seymour Publications, 1989.

Seymour, D., and J. Britton. *An Introduction to Tessellations.* Palo Alto, CA: Dale Seymour Publications, 1989.

Silvey, L, and L. Taylor. *Paper and Scissors Polygons and More.* Palo Alto, CA: Dale Seymour Publications, 1997.